I0055998

Digging Earth

Extractivism and Resistance on Indigenous Lands of the Americas

Edited by

Catherine Bernard

Digging Earth: Extractivism and Resistance on Indigenous Lands of the Americas

Edited by Catherine Bernard

This book first published 2024

Ethics International Press Ltd, UK

British Library Cataloguing in Publication Data

A catalogue record for this book is available from the British Library

Copyright © 2024 by Ethics International Press

All rights for this book reserved. No part of this book may be reproduced, stored in a retrieval system, or transmitted, in any form or by any means, electronic, mechanical photocopying, recording or otherwise, without the prior permission of the copyright owner.

Print Book ISBN: 978-1-80441-068-4

eBook ISBN: 978-1-80441-069-1

Contents

Part 3

Taking Back Indigenous Lands, Water and Epistemes

Contributors

Catherine Bernard is Professor of art history at SUNY Old Westbury and obtained her Doctorat d'Etat from the Sorbonne University. She has written extensively on Diaspora, post-colonial and contemporary art. Her work has been published in African Arts (UCLA); Parkett Magazine; The Art Journal (College Art Association); Documents of Contemporary Art, (Whitechapel Art Gallery and MIT Press); Nka Journal (Duke and Cornell Universities); Les Carnets du Bal, Paris; the Blaffer art Museum, Houston. Her curatorial work includes more than 20 exhibitions and several catalogues on contemporary artists for the Studio Museum in Harlem, the Neuberger Museum of Art; Wallace Gallery, SUNY Old Westbury; Hunter College Galleries, CUNY; the Katonah Museum of Art; Museo Gurvich, Montevideo. *Trees Also Speak*, 2018, showcased the work of contemporary Indigenous American artists and received a grant from the National Endowment for the Humanities.

Susanne Berthier-Foglar is Professor Emerita at University Grenoble Alps, France. She has been a partner and participant in two European EIT Raw Materials projects: *Mine Heritage* with an emphasis on the history of mining, and *Open Your Mine* a field course for Master students of participating universities where aspiring professionals acquired experience on the meaning of land for Indigenous Peoples. She published a monograph on the Pueblo of New Mexico: *Les Indiens Pueblo du Nouveau-Mexique*, Presses Universitaires de Bordeaux, 2010. On the resources of the Americas she co-edited *Ressources Minières dans les Amériques* (with S. Tolazzi and F. Gaudichaud), IdeAs Nr8, 2016-2017. On genetic versus cultural identities she co-edited *Biomapping or Biocolonizing* (with S. Collignon-Whittick, and S. Tolazzi), Rodopi, 2012. On borders, real and metaphorical, she co-edited (with Paul Otto) *Permeable Borders*, Berghahn, 2020, and *Migrations and Borders in the United States: Discourses, Representations, Imaginary Context*, Représentations, CEMRA, 2018. Other publications are listed in ORCID.

Antonia Carcelén-Estrada is an activist, translator, and scholar of comparative literature, cultural race studies, oral history, and early-modern and medieval studies. She has worked at the University of

Massachusetts, Amherst, as well as for the College of the Holy Cross, Universidad San Francisco de Quito, and Sarah Lawrence College. She is currently on a scholarship at the University of New Mexico's Latin American and Iberian Institute carrying out comparative research on the double Spanish-English colonization of Pueblo nations. Her publications on Indigenous intercultural translation include, *Zapatista Stories for Dreaming An-Other World* (2022), "Oral Literature" (2018), "Translation and Activism" (2018), "Weaving Abya-Yala" (2017), "What does *Sumak Kawsay* Mean?" (2016), "Rewriting Memory" (2012), and "Covert and Overt Ideologies in the Translation of the Wycliffe Bible into Huao Terero" (2010). Other decolonial research includes, "Oral Histories in the Black Pacific" (2022), "Decolonizing Oral History" (2021), and "Jewish and Islamic Foundations" (2020).

Carolina Caycedo (London,1978) is a Colombian multidisciplinary artist living in Los Angeles. Her immense geographic photographs, lively artist's books, hanging sculptures, performances, films, and installations are not merely art objects but gateways into larger discussions about how we treat each other and the world around us. Through her studio practice and fieldwork with communities impacted by large-scale infrastructure and other extraction projects, she invites viewers to consider the unsustainable pace of growth under capitalism and how we might embrace resistance and solidarity. Process and participation are central to Caycedo's practice, she contributes to the reconstruction of environmental and historical memory as a fundamental space for climate and social justice. Caycedo is a nominee for the 10 Artes Mundi prize in Wales, a 2023 United States Artists Fellow, and a 2023/24 Artist in Residence at the Getty Research Institute.

Jeffrey De Blois is Associate Curator and Publications Manager at the Institute of Contemporary Art/Boston, where he has organized *Tammy Nguyen; Rose B. Simpson: Legacies; Napoleon Jones-Henderson: I Am As I Am – A Man; Raúl de Nieves: The Treasure House of Memory; Eva LeWitt; Carolina Caycedo: Cosmotarrayas;* and *Caitlin Keogh: Blank Melody,* among others. He has been a critical contributor to several other projects at the ICA, including *To Begin Again: Artists and Childhood, Sterling Ruby, Less Is a Bore: Maximalist Art & Design, William Forsythe: Choreographic Objects,* and *Art in* *the Age of the Internet, 1989 to Today*. He has written essays on the work of artists Marlon Forrester, Carolina Caycedo, William Kentridge, Caitlin Keogh, and Sterling Ruby. Before joining the ICA, De Blois was curatorial fellow at MIT List Visual Arts Center. He holds a MA from Boston University in the History of Art & Architecture.

Jeremy Dennis (b. 1990) is a contemporary fine art photographer, an enrolled Tribal Member of the Shinnecock Indian Nation in Southampton, NY, and lead artist and founder of the non-profit Ma's House & BIPOC Art Studio, Inc. In his work, he explores Indigenous identity, culture, and assimilation. Dennis holds an MFA from Pennsylvania State University, State College, PA, and a BA in Studio Art from Stony Brook University, NY. He currently lives and works in Southampton, New York on the Shinnecock Indian Reservation.

DesertArtLAB is an interdisciplinary environmental arts collaborative co-directed by April Bojorquez and Matt Garcia. Their work promotes Indigenous/Chicanx perspectives on ecological practice, food sovereignty, self-determination, and climate change. DesertArtLAB's projects activate public space through participatory artworks and support the restoration of desert environments and their foodways through zero irrigation regrowth projects. DesertArtLAB have presented their work nationally and internationally at Ecole Nationale Supérieure des Beaux-Arts, Paris, France; The Dom Museum, Vienna, Austria; The Museum of Contemporary Native Art, Santa Fe, NM; the Museum of Contemporary Art Santa Barbara, Galería de la Raza, San Francisco, CA, among many others. April and Matt are recipients of the Creative Capital award and were 2021 Mellon Artists in Residence at the Colorado College Fine Arts Center Museum. Bojorquez and Garcia live in Pueblo, Colorado where they work as artists/educators. Garcia is an Assistant Professor of Creativity + Practice at Colorado State University Pueblo.

Elizabeth "Betsy" S. Hawley is an art historian, writer, and curator specializing in art of the Americas and modern and contemporary art. Her research often focuses on twentieth and twenty-first century Native North American art, and other areas of expertise include feminist/women's art, activist art, ecocritical art, and art of the American West. Her work has been supported by the Lunder Institute, Wolfsonian-FIU, and Pittsburgh Foundation. Hawley is an assistant professor of art history at the University of South Alabama and an independent curator. Recent exhibitions include *Landscapes of Survivance* at the Santa Clara University Art Gallery and *Borderwaters* at the Alabama Contemporary Art Center.

Lynn Holland teaches US-Latin America Relations at the Josef Korbel School of International Studies at the University of Denver. Her research is focused on environmental activism and land rights issues related to mining and other forms of extractivism, drug trafficking, immigration policy, and state-society relations in southern Mexico, Central America, and Peru. She has taken students to Chiapas, Mexico to study the impact of large-scale development projects and the sustainable alternatives that are emerging in the region and has served as an expert witness for asylum seekers coming from Mexico and Honduras.

Erin Joyce is a scholar of contemporary art and has organized over 35 solo and group exhibitions including *Between Beauty and Decay* (Artspace New Haven, 2017), *Still Life No. 3: Raven Chacon* (Heard Museum 2019), *Erika Harrsch: Moving in the Borderlands* (Idyllwild Arts Foundation, 2022), and *Crafting Resistance* (Arizona State University Art Museum, 2023). Joyce is a frequent contributor to *Hyperallergic*, and has had writing featured in *Salon, Selvedge Magazine, Canvas Magazine,* and *Native American Art Magazine*. She is a 2023 winner of the Rabkin Prize for arts journalism from The Dorothea and Leo Rabkin Foundation. Joyce's work has garnered attention from publications including *Vogue Magazine, The New York Times, The Economist, The Art Newspaper,* and *Forbes*. She has lectured at venues such as the Yale Center for British Art (New Haven, Connecticut), the School for Advanced Research (Santa Fe, New Mexico), and Fire Station Artist Studios (Dublin, Ireland). Joyce holds a BA in the History of Art from the University of North Texas and MA in Museum Studies from Johns Hopkins University.

Dr. Kelsey Leonard holds a Canada Research Chair in Indigenous Waters, Climate and Sustainability and is an Assistant Professor in the School of Environment, Resources, and Sustainability in the Faculty of Environment at the University of Waterloo, where her research focuses on Indigenous water justice and its climatic, territorial, and governance underpinnings. As a water scientist and legal scholar, Dr. Leonard seeks to establish Indigenous traditions of water conservation as the foundation for international water policymaking. She represents the Shinnecock Indian Nation on the Mid-Atlantic Committee on the Ocean, which is charged with protecting America's ocean ecosystems and coastlines. She also serves as a member of the Great Lakes Water Quality Board of the International Joint Commission. She is an enrolled citizen of Shinnecock Indian Nation.

Juan Pablo Pacheco Bejarano (Bogotá, 1991) is an artist, writer and educator driven by the interrelations between technology and ecology, seeking to imagine media and mediations beyond extraction. Through texts, collaborative spaces, audiovisual projects and installations, his research dives into the fractal relations between technological infrastructures, fermentation, water and telepathy. Juan Pablo has been a cultural programmer at Plataforma Bogotá, a lab for art, science and technology, and at Espacio Odeón, a contemporary art space. He has also taught at the Javeriana and Andes universities in Bogotá, and at the Royal Academy of Arts The Hague. His work has been disseminated and developed at multiple cultural centers and institutions around the world.

Nicole Sault draws on the perspectives of ethno-ornithology and anthropology in working with peoples of México, Costa Rica, Perú, and the U.S. This work honors the voices of Indigenous elders with traditional environmental knowledge (TEK) who are resisting ongoing neo-colonial land theft and repression. In defending the land and water she calls attention to the role of bird teachers like condors and hummingbirds. Topics she addresses include kinship and godmothers, human rights and torture, and also chiles and the temazcal. This research and writing are in Spanish as well as English, and she has taught at the Universidad de Costa Rica and Santa Clara University in California. She earned a doctorate from the University of California at Los Angeles, and her publications include *Many Mirrors: Body Image and Social Relations*, Rutgers University Press. www.sallyglean.org/sault/

Kaila T. Schedeen, PhD is a curator and art historian based in Austin, TX. She received her PhD from the University of Texas at Austin in 2023 and her MA from the University of Delaware in 2016. Kaila specializes in contemporary American art with a particular focus on Indigenous artists and artists of the African Diaspora who critically examine the terms of identity, belonging, and nationhood in the United States through photography, performance, and multimedia works. Her writing has been published by University of Delaware Press, Rutgers Art Review, Grove Art Online, and Sightlines Magazine, and she has held positions at The Harry Ransom Center, UT; the Visual Arts Center, UT; Fred Jones Jr. Museum of Art, University of Oklahoma; John L. Warfield Center for African and African American Studies, UT; Winterthur Museum, Garden and Library; and Mechanical Hall Gallery at the University of Delaware.

Morten Søndergaard is an internationally acclaimed Curator and Associate Professor of Media Art. With a background as Curator/Deputy Director at the Museum of Contemporary Art in Roskilde (1998-2008), he is academic director and co-founder of the Erasmus master's in media arts Cultures program (since 2015). Co-founder of the POM – Politics of Machines Series (with Laura Beloff, Aalto University, since 2017) and co-found of the ISACS – Sound Art Curating Conference Series (w Peter Weibel, ZKM, since 2013). He has published internationally on media art, sound art, curation, and the digital archive experience. He was Contributing Editor at LEA – Leonardo Electronic Almanac (MIT Presse, 2011–14), and Editor of MEDIEKULTUR Journal (2013-15). Guest editor at Leonardo Music Journal (MIT Press, 2019-2021). Morten Søndergaard is the Curator of several international exhibitions, including MAGNET (2007) and The Unheard Avant-gardes (2012) at ZKM, Karlsruhe.

Edward Charles Valandra is Síčáŋǧu Thithuŋwaŋ and a citizen of Očhéthi Šakówiŋ Oyáte. He was born and raised in his settler-occupied homeland, the Očhéthi Šakówiŋ Oyáte Makȟóčhe. Dr. Valandra has served his people in various capacities: a legislator, and a senior administrator in his nation's K12 chartered school. He is currently a member of the 1894 Sioux Nation Treaty Council. He received his BA Minnesota State University-Mankato, MA from the University of Colorado-Boulder, and PhD from SUNY-Buffalo. His research focuses on the national revitalization and liberation of his nation and of other communities of color. Dr. Valandra's current role is Senior Editor at Living Justice Press. He is the editor of *Colorizing*

Restorative Justice: Voicing Our Realities and author of *Not Without Our Consent: Lakota Resistance to Termination, 1950-1959*, and it currently the editor of *Colorizing Circle Practices: Naming the Silences* book project.

Will Wilson is a photographer and trans-customary artist who spent his formative years living on the Navajo Nation. His photography practice centers around the continuation and transformation of customary Indigenous cultural practice, countering the 'archival impulse' embedded within the historical imaginaries of white settler colonialism. Through various methods of photography, Wilson combines digital technology, historic photographic processes, performance, and installation around themes of environmental activism, the impacts of cultural and environmental change on Indigenous peoples, and the possibility of cultural survival and renewal. Wilson studied photography, sculpture, and art history at the University of New Mexico where he received his Master's in Fine Arts. Wilson has been honored with the Eiteljorg Native American Fine Art Fellowship, Joan Mitchell Foundation Award for Sculpture, Pollock-Krasner Foundation Grant for Photography and he was the Doran Artist in Residence at the Yale University Art Gallery. Wilson has held visiting professorships at the Institute of American Indian Arts, Oberlin College, and the University of Arizona. His work is exhibited and collected internationally and he is Associate Professor of Photography at University of Texas, Austin.

Rachel A. Zimmerman is assistant professor of art history at Colorado State University Pueblo. She earned her bachelor's degrees at New Mexico State University and her MA and Ph.D. at the University of Delaware. Her primary area of research is art and material culture in the early modern Portuguese empire, and she held a fellowship at Winterthur for her research on hammocks, ceramics, and textiles in eighteenth-century Brazil. She is an editor for Smarthistory.org and contributed a chapter to their open-access online textbook Reframing Art History. Since 2022, she has served as a Faculty Fellow at CSU Pueblo to help her colleagues with pedagogical opportunities and challenges such as online teaching, alternative grading methods, supporting disabled students, and teaching in the time of generative AI.

Acknowledgements

Digging Earth: Extractivism and Resistance on Indigenous Lands of the Americas is the result of a collaboration with the writers and artists in this volume who generously shared their work and insights on the subjects of extractivism and the resistance struggles in many Indigenous communities of the American continent. I am grateful that they accepted to donate their work and for their willingness to answer many questions and share their knowledge with me. Your scholarship and creativity give this book its value.

I am grateful to the artists who have authorized the reproduction of their works: Pia Arke and the Pia Arke Foundation, Carolina Caycedo, Jeremy Dennis, DesertArtLAB (Matt Garcia and April Bojorquez), Juan Downey, Courtney Leonard, Emma Robbins, Barbara Santos, Angelica Trimble-Yanu and Will Wilson. Your artworks establish loud and clear the critical impact of visual narratives to understand the histories of Indigenous communities and their homelands.

Sarah Palmer and Ben Williams at Ethics Press have helped me every step of the way. Thank you for your patience and timely responses to my questions always and for easing the process of putting this book together.

The Hopi at First, Second and Third Mesas who let me stay during ceremonial dances and prayers and have welcomed me in their houses to share a meal, sparked my will to learn about Indigenous lands and communities. Janice and Joseph Day at Tsakurshovi on Second Mesa gave me sound advice, warnings and proper behavior etiquette, while marking on a map of the Four Corners trails to places I should go see.

I give my profound gratitude to all of them and to the many individuals I met during my travels who helped me with directions, both physical and intangible.

A number of friends and colleagues have sustained this research and brought ideas, suggestions and support. I am grateful to Barbara Frank,

Johan Grimonprez and Bruno Ricard for endlessly listening, providing support, suggestions and good humor. Thank you to Thierry Pérémarti for suggesting: Digging Earth, originally as title of a conference paper.

Last to Milo, my son and travel companion and to his generation, I hope this book may provide a small sparkle to pursue and learn from the lessons we have not heeded soon enough.

May the work done in this book benefit the humans, other-than-humans, waters, plants and rocks of the earth.

Catherine Bernard, August 2023

Introduction
Rumbling Lands and Muttering Rivers

Catherine Bernard

Digging Earth: Extractivism and Resistance on Indigenous Lands of the Americas is a collection of essays and artists' contributions that examine resource and epistemic extractivist practices on Indigenous homelands throughout the American continent and the resistance movements led by Indigenous leaders, activists, scholars and artists fighting them. Resource extractivism includes the exploitation of minerals and ores, the use of millions of gallons of water for cleaning and transporting them, the construction of dams to produce hydroelectric power, logging, deforestation, and oil drilling, solely driven by profit. The authors and artists in *Digging Earth* denounce the predatory extractivism of neocolonial corporations and their impact on the environment of local and Indigenous communities; they frame these activities within the settler colonialist ideology that converts the natural environment into a capitalist resource, ready to be plundered; they address topics of Indigenous sovereignty, environmental justice, and land's rights; they write about the power and dissent in the communities who live in regions of extraction.

In addition to resource extraction, the extraction of Indigenous epistemes and ontologies was instrumentalized during the nineteenth century and beyond by anthropology and ethnology, and they became essential tools to legitimize the colonization process. Today, the extraction of Indigenous knowledge systems is practiced by several government agencies and international institutions as they consider traditional Indigenous knowledge for their climate change strategies, ready to be used as models for new environmental ethics, or solutions to the environmental crisis generated by resource extraction. Not only is this assumption problematic and is at best a form of wishful thinking, but it also avoids asking who would benefit most from the appropriation of this knowledge.

As we, humans and non-humans, hasten toward a foretold collective ecocide, contributors to the volume point to the impacts of the climate crisis being felt by Indigenous nations and countries in the Global South, having been affected by the ecologically destructive consequences of extractivism for many decades and who are first in the line of destruction.[1] Environmental justice can only be understood from the point of view of the south-north divide that has resulted in the forced migrations of millions, increased pollution and climate related disasters, and the systemic destruction of ecosystems throughout the Global South.

The collective works in this volume respond to the dead end of extractive capitalism with stories of dissent and possibilities and by creating artworks that imagine futures and alternative models, new strategies of thinking, and struggles that dismantle the extractivist paradigm as a social and economic template. Generated by artists, activists, writers and cultural producers, these models envision shared spaces of cohabitation and relation between human, other-than-human species, and the natural environment.

How to Speak and Stand in Solidarity?

While researching Indigenous arts and histories for the past two decades and learning about the multiple histories of the Four Corners region, the various issues surrounding mining came to the fore, and it became clear to me that accounting for such complexity was a daunting and complicated task. To speak and stand in solidarity, was it enough for me to document and write about mining and the threats faced by Öngtupqa (the Hopi name for Grand Canyon) and the multiple nations—Hulapai, Havasupai, Diné, Hopi, Mountain Ute, Ute—who live near Öngtupqa, draw water and subsistence from the Colorado River winding in its depths, and whose histories and cultures are woven in its geological time? Or to report the threats on the religious practices embedded in its waters and rocks, for example, those of the Hopi for whom the Place of Emergence of the world is located at the confluence of the Colorado River and the Little Colorado River, at the mouth of Öngtupqa?

An example provides some focus into the resistance undertaken by Indigenous Peoples living near Öngtupqa. A 1986 enterprise, Canyon Mine, was recently resurrected and renamed the Pinyon Plain Mine. Energy Fuels Resources, the present owner, was in the process of obtaining permission to renew uranium extraction just miles away from the entrance to the Grand Canyon National Park, targeting a uranium deposit located 1,400 feet below the surface. The intertribal Red Butte Gathering in October 2018, organized by the Havasupai protested the mining project. Following the denial of a 2018 petition filed by the Havasupai to the US Court of Appeals, the Havasupai Tribal Council sent a formal letter of opposition to the Arizona Department of Environmental Quality in 2022, reiterating its opposition to the Pinyon Plain Mine, requesting consultation and coordination with the Arizona Department of Environmental Quality to recognize the value of tribal traditional knowledge, as well as demanding that basic data on the mine's operations and contaminants be shared. A decision regarding the start of mining in Pinyon Plain Mine was pending as of 2022 (Grand Canyon Trust, 2022), until the recent designation of Baaj Nwaavjo I'tah Kukveni as a National Monument by the Biden administration banned uranium mining within its limits. This decision could not have happened but for the unified opposition of the various nations involved in resisting the project: Havasupai, Hopi, Hualapai, Paiute, Navajo, Yavapai-Apache, Zuni and Colorado River Indian Tribes.[2] However, the fact that that other similar operations continue to dot the Four Corners regions, the homelands to many Indigenous nations, should not be ignored.

This brief account describes one of many attempts at exploiting the land and threatening the environment and health of the inhabitants of the region. Similar situations of resource exploitation and disregard for the communities they affect are ubiquitous throughout North, Central, and South America. As a result, the question of extractivism became central to my research. Some important issues needed to be examined. How social and political power structures that are fundamentally unequal would frame this research: who retains and distributes the knowledge accumulated through research and publishing? As a European scholar and curator engaged in academic work, speaking from the United States, where do I stand in relation to the colonial context defining these threats? As part of the institutional academic structure, I needed to examine my situation in relation to the production of knowledge in a settler colonialist context. Additionally, what role did the tropes of the settler imagination fueled through the prism of popular culture (Disney, Hollywood, Netflix) and news media play in my choices? To learn how to speak with and stand with, instead of to speak for and stand for, one must ask these questions. There were a few possible signposts.

Some of my counterparts, Anglo-American and Europeans scholars who write about Indigenous cultures and Indigenous Peoples have adopted the term *ally* to define their position in the ongoing context of settler colonialism. Could this notion help to account for my own relation to the political, social and theoretical frames of my research? I kept coming back to the words of Jaskiran Dhillon from her essay "Notes on Becoming a Comrade: Indigenous Women, Leadership and Movement(s) for Decolonization": "Benign notions of allyship, or solidarity with no teeth, has been rightfully critiqued for reinscribing colonial relations of domination, and doing little to interrupt or dismantle white settler power and broader colonial structures upholding white supremacy" (Dhillon, 2019, p. 43).

Another signpost was offered by historian Nick Estes in *Our History is the Future,* where he casts Indigenous resistance in the historical and generational frame of Indigenous struggles since the arrival of settlers on Turtle Island. In Estes's work and that of other Indigenous scholars such as Edward Valandra (chapter 5 in this volume), being *a good relative* is the foundation of such historical movements: "For the Oceti Sakowin, the affirmation Mni Wiconi, 'water is life,' relates to Wotakuye, or 'being a good relative.' Indigenous resistance to the trespass of settlers, pipelines, and dams is part of being a good relative to the water, land, and animals, not to mention the human world" (Estes, 2019, p. 21). Pollution is entitled with more rights than the lands and the peoples, as it trespasses unhindered through the aquifers, the air, and the soil while Indigenous lands are parcellated and legislated through settlers' politics.[3] Learning to understand the natural world as kin is another aspect of speaking and standing in solidarity.

Furthermore, to speak and stand in solidarity demands being critical of scholarship that represents Indigenous cultures as a general paradigm, describing nonspecific cultural or political groups. This approach is denounced by Indigenous scholars such as Jennifer Nez Denetdale, Winona LaDuke, Simon J. Ortiz, Dana Powell, Melanie Yazzie, and others, as they claim that any research addressing topics of Indigenous political struggles requires that they are understood within specific historical, political, and cultural frames. Melanie Yazzie, in "Decolonizing Development in Diné Bikeyah" (Yazzie, 2018), argues that an analysis of the models of imperialist and colonial capitalist development must be central to discussions of decolonization and Indigenous cultures. She also asserts that it is fundamental to recognize how each situation generates specific struggles and narratives that are inscribed within their own spatial temporalities. Only then can the colonial settler structures be accounted for and deconstructed.

In the present context of settler colonialism, acknowledging my own position as being part of the dominant settler class and examining it in regard to the history of appropriation, exploitation, and genocides linked to colonial destructive powers is unavoidable. To be a political supporter becomes possible only by learning from the work accomplished by Indigenous scholars and artists, from their ongoing experiences, narratives, engagements and political actions. This acknowledgement needs also to be framed with questions about my own social and political history; to examine the working relation between the diverse actors contributing to the project—authors, artists, and editor, and to establish a degree of accountability about who retains power and knowledge. These questions have guided me through this project.

The contributors in *Digging Earth* critically deconstruct the extractive paradigm. Each of them, Indigenous and non-Indigenous writers and artists acknowledge their own position and unique perspective and underline the critical agency of specific cultural and political webs, intertwined with the voices of distinct Indigenous communities. Their work contributes to a scholarship that aims at speaking and standing with Indigenous Peoples in their resistance movements.

Extraction and Settler Colonialism on Indigenous Homelands

Extractivism must be understood within the history of settler colonialism. It started as soon as the first occupation of the land was achieved after the first cross Atlantic European expedition. As Alberto Acosta states: "Extractivism is a mode of accumulation that started to be established on a massive scale five hundred years ago. The world economy—the capitalist system—began to be structured with the conquest and colonisation of the Americas, Africa and Asia" (Acosta, 2013, p. 62). The histories, cultures and economies of the American continent were radically altered by the appropriation of Indigenous lands by European colonizers. The ensuing invasion was driven by territorial expansion, the exploitation of resources, and the elimination of Indigenous Peoples who stood in the way of the settlers. This process continues today via the extractivist politics of the neocolonial corporations that have replaced the old colonial nation-states while using similar tactics.

The name *America* first appeared on a 1507 map titled: *Universalis Cosmographia* based on travel accounts by an Italian merchant named Amerigo Vespucci, who worked for the republic of Florence and traveled across the Atlantic in the late fifteenth and early sixteenth centuries. It is speculated that the first of his travels took him to the Gulf of Mexico. The next time he sailed across the Atlantic, he went as far as the Amazon and on his return sailed through the Caribbean. These expeditions were financed by Spain. Soon, Spain was the largest of the colonial empires in the American continent, followed by Portugal, and at the dawn of the seventeenth century, by France, England, and the Netherlands. The full-scale takeover and exploitation of Indigenous Peoples and their homelands on the continent had started.

In 1540, after traveling with his army through what is now the state of Arizona, Francisco Vazquez de Coronado's expedition arrived in what is now New Mexico. The expedition was based on the rumored existence of the seven cities of Cibola, a legend connected to that of El Dorado, the fabled land sought after by Spanish armies during the sixteenth century throughout the northern part of South America (today's Colombia, Venezuela, and the Amazonia). In spite of high hopes, Coronado only found the adobe towns of the ancient Puebloans, such as the villages of

Zuni and Tiwa peoples who organized wars of resistance against the advance of the Spaniards yet couldn't stop them. In lieu of gold, Coronado discovered rich turquoise and copper reserves and soon after, the Spaniards and their missionaries occupied the land of the Puebloans, mining and attempting the subjugation and conversion of the inhabitants, while bringing cattle and their agricultural techniques with them.

This short account clearly contextualizes the arrival of the first Europeans on Turtle Island with the search for minerals and the takeover of the lands of the Puebloans. On the Acoma homeland in 1599, strong resistance was led by Zutapacan and the Acoma people who fought against the conversion to Catholicism, religious persecution, and the rule of *encomiendas*, the right conferred by the Spanish crown to its colonies and used by Juan de Onate and his army to impose forced labor on Indigenous Peoples. Onate was accompanied by five hundred settlers and thousands of heads of livestock in his attempt to gain control of the region. The battle ended with a Spanish victory and the massacre of Acoma people. Other resistance movements continued to be fought by the people in the southwest. The Pueblo uprising of 1680 unified the Puebloans under the leadership of Po'pay of Ohkay Owingeh (San Juan Pueblo). Puebloans and their Diné, Hopi, and Apache allies captured Santa Fe, overthrew the Spanish rulers, and established their sovereignty for twelve years. The Franciscan missions were destroyed, and the colonists fled south to what is now Mexico. Pueblo historians call the insurrection the first American revolution, challenging historical erasure and claiming agency through an Indigenous lens.

Art as Part of the Colonial Apparatus

In addition to essays by scholars, *Digging Earth: Extractivism and Resistance on Indigenous Lands of the Americas* includes the contributions of four artists who share portfolios of their work accompanied by critical essays. The works by Carolina Caycedo, Jeremy Dennis, the collective DesertArtLAB, and Will Wilson propose a visual epistemological critique of the appropriation of Indigenous lands, and expose the blatant imperialism of resource exploitation. Through sculpture, performance, video, and photography, these artists claim aesthetics as an instrument of transformation in the sociopolitical spaces of Indigenous homelands and remind the viewer of the fundamental ontological role that water, mountains, and plants play in the histories and lives of their communities.

It seems pertinent here to take a look at the role played by cultural and aesthetic systems in the history of the colonization of the Americas and the ideologies that commanded settlers to take over whatever and whomever stood in their trajectory, including the elimination of Indigenous Peoples. White settlement could only succeed that way. One major tool at the disposal of the settlers was the Doctrine of Discovery, issued in May 1493 by Pope Alexander VI, the foundational religious and cultural document sustaining colonization on a global scale. It supported Spain and other colonial forces, justifying the European takeover of Indigenous lands and Peoples. Any land unoccupied by Christians was considered *Terra Nullius* and could be claimed as discovered, invalidating any Indigenous preexisting governments, structures, or land sovereignty.

During the Renaissance period, colonial ideology was developed in parallel with a particular visual system enabling the rationalization of the pictorial space. Linear perspective (*De Pictura* by Alberti, 1435) allowed artists to quantify space and propose a "scientific" vision of the world. It became used by the most important painters of the era who were supported by the wealthy merchants and the political powers funding the colonial expeditions. Linear perspective provided a mathematical system to organize the surface, starting with defining a point on the surface: the vanishing point. Aptly named, the vanishing point became the place on a geometric grid where lines converged, allowing space to be measured and infinity visualized. Soon, it dominated European art, and the few artists who rebelled against this system were persecuted by the tribunals of the Inquisition (Lepenies, 2013).

It seems fitting, therefore, that the logic of mastering—conquering—the pictorial space and making it a mimicry of reality, would find an echo in the development of faraway expeditions, accompanying the ideology of conquest. The Western imaginary was ready and armed with the potential for exploring the infinite and eventually conquer and dominate it.

Peoples of Asia, Africa, Australia, and the Americas living in vast lands had developed their own specific spaces of representation, that rest on depictions relating to specific meanings, beliefs, and cultural, political, and social narratives, at time sacred narratives. These models rejected the rationalization of space and the hierarchy it implies in terms of scale or depth. The arts of Indian, Chinese, Japanese, African peoples, and many others showed no interest in reproducing reality as Europeans did; rather, they understood art as envisioning and transforming reality rather than copying it. These aesthetic systems became the key to renewing an exhausted Realism in Western art at the end of the nineteenth century when the Modernist era appropriated the canons of representation of artists in Africa, Asia, and other lands to sustain their quest for renewal. The complexity of these cultural dynamics is entrenched in the ideologies of colonization and the extraction of land and knowledge that have dominated the West since the fifteenth century.

Another example of the entanglement of colonization with artistic practices can be found in nineteenth century landscape paintings in the United States. The development of what is generally considered in North America as the first iteration of a "true" American art (notwithstanding the millennia of artistic productions of Indigenous arts), is known as the Hudson River School, since most of the painters started their careers in the region of what is now New York State. The Hudson River School became closely related to the expansion of settler colonialism. Large landscape paintings, representing vistas of yet uncolonized lands, decorated the Northeastern mansions of rich landowners, for whom the paintings, along with richly decorated furniture, stood as a testimony asserting their status and power. The ruling elite commissioned paintings from the artists of the Hudson River School, works that supported Manifest Destiny and the 'God-given right' to settler expansion into Western territories. Magnificent idealized landscapes, dotted with golden sunsets and light, achieved with the technique known as Luminism, created idyllic representations of wilderness, ready for plunder for those who would dare venture there.

Some of these paintings (Bierstadt's *Lander's Peak* for example) included the depiction of peaceful Indigenous ways of life, at a time when Indigenous populations were decimated by the settlers and the military. The nostalgic take on the actual situation, subsumed its reality, while the construction of railroads and the progression of settlers were destroying the lands, villages and lives of the Indigenous communities standing in their way. It was thus better to represent them as remnants of an idyllic past, better suited to museums and dioramas—yet another way to extract Indigenous Peoples out of history.

These examples highlight the role of cultural tropes in the manifestation and achievement of settler colonialism and its expansionist ideology and should not be underestimated when discussing the politics and economics of extractivism. After all, the Doctrine of Discovery, first supporting the colonizers' claim to Indigenous homelands, was only rescinded by Pope Francis in March 2023, while this book was taking shape.

The Violence of Extraction: Minerals, Water, Bodies, Ontologies

Resource Extraction

Extractivism is commonly understood as resource extractivism in the context of a global-scale exploitation of minerals mostly benefiting the economies of the Northern hemisphere as well as some of the "emerging markets" of the ancient European colonies. Some of these minerals are used to power new green technologies, also dependent on the extractive process. Extractivism mostly benefits the service industries and new technologies, improving the living conditions in the Global North, and resting on cheap labor and raw materials extracted in the Global South. This hegemonic capitalist mode of production is achieved through the exploitation, mainly by transnational corporations, of Indigenous lands and resources, forcing the displacement of entire communities, destroying the environment, and fostering social divisions and cultural erasure. Indigenous Peoples agree to work for mining companies in spite of low salaries and the attendant risks to their health (which are seldom disclosed), in large part because the current occupation of Indigenous homelands include few other economic alternatives besides tourism (another exploitative economy) and because mining profits thrive on the continuity of a capitalist logic of class inequality.

The US government and the mining companies operating in Diné Bikeyah, the Diné homeland, didn't disclose the health risks associated with uranium to the thousands Diné miners who worked for minimal wages in the uranium mines on their homeland from the 1940s to the 1980s, the results of which are still visible today, as the health of Diné people continues to be affected by the proximity of uranium deposits and leaks in the water from the more than five hundred abandoned mines that still have to be closed (see chapters 2 , 6, and 9 in this volume).

Similarly, the construction of enormous dams throughout the Americas affects a large number of populations, yet are planned and executed without consultation with riverine communities. An estimated 5.7 million people in South America are negatively affected by these projects. Villages and lands are flooded, and people are forcibly displaced. Energy and construction corporations, invested in the building of dams, offer low wage jobs to those who lose their entire way of life, their livelihood, and do not hesitate to resort to violence if any of the communities affected by the destruction brought by these projects bring attempt to resist (Grassroots International, 2017).

The hegemony of development models based on extractivism proposed by the West has succeeded in extending capitalist social and political relations globally, so they come to govern the lives of millions. Characterized by exploitation, and domination of the lands, people, and other species inhabiting them, Western corporations refuse to acknowledge world views that highlight interspecies and ecological interdependencies and co-responsibilities. If such recognition were to happen, it would necessarily also mean dismantling models that rest on class separation, racial and ethnic exploitation, and the division between humans and other species.

Extracting Bodies — Gendered Violence

Extractivism rests on violence against the natural environment, humans, and other-than-human beings. Exerted against Indigenous bodies, such violence is inseparable from the history of settler colonialism. Between 1819 and 1969, the US Federal Indian Boarding School system forcibly removed hundreds of thousands of children (this number is only an estimation) from their families and placed them in Indian boarding schools (407 in the US alone). Forced assimilation subjected children to abuse, mistreatment, child labor, health hazards, and death (Newland, 2022). In Canada, for more than

150 years, First Nations, Inuit, and Metis nations children were subjected to similar treatments. The staggering generational trauma that resulted has just started to be accounted for and acknowledged by the US and Canadian governments. In Central and South America, policies of forced removal, violence, and assimilation that coerced Indigenous children in mostly Christian missions' residential schools from the 1800s until 1970 are seldom acknowledged by politicians and governments.

In today's settler colonial lands, a particularly vicious aspect of this violence is found in the extraction of the bodies of Indigenous women, girls, and Two Spirit peoples. Feminicide against Indigenous women in their role of caretakers of the land, resources, and animals and as community organizers has been rampant since the beginning of colonization centuries ago (see chapter 6 and 7 in this volume). They are central voices that drive resistance in the fight against the appropriation of water and resources in many Indigenous homelands: Water Defenders in Peru; women of African descent and Indigenous women in the Ecuadorian and Colombian Chocó; Mujeres Amazonicas; Diné women in the Southwestern United States; Women of the Oceti Sakowin against the Dakota pipeline and many others, are formidable presences in the struggles against resource extractivism.

As Melanie Yazzie explains: "These Indigenous women land defenders also point out that the violence of resource extraction affects not only the lands that are plundered and pillaged during resources raids . . . but also the bodies of Indigenous people—and women, youth, and LGBTQ relatives in particular. This land/body relationality is bound by and through an intergenerational toxicity caused by industrial pollution, often as a result of resource extraction" (Yazzie, 2018, p. 33).

In spite and because of their powerful stances against exploitation near mines, pipelines and endangered water resources, Indigenous women and girls are attacked and murdered at a high rate. Sexual violence against them occurs in alarming numbers near communities where extractive industries are located. The community based grassroot movement MMWIG2—Missing and Murdered Indigenous Women, Girls and Two Spirit people (and their red hand symbol), raises awareness of the violence against Indigenous women and LGBTQ people in North America. In the United States and Canada, abuse, rape, and murder statistics are staggering:

Indigenous women are murdered at a rate ten times higher than the national average and murder is a leading cause of death among young Indigenous women (Urban Indian Health Institute, 2018). The lack of media coverage in these cases helps conceal the gravity of the issue. Extractive industries of mining, logging, and oil and coal industries are heavily implicated in the violence against Indigenous women. They bring an influx of transient male workers to rural areas, often near or on Indigenous homelands. The violence rate near the "men camps" is much higher than any other places in states that have extractive industries (Stern, 2021). In extractive regions throughout North America Indigenous women, girls, and Two Spirit people are viewed as commodities. This violence exerted against them is spread throughout the continent. Women of African descent in Ecuador and Colombia, and particularly those living in the Esmeraldas are murdered in great numbers in regions where mining, industrial farming, and deforestation are the rule (see chapter 8 in this volume). The violence against women exists wherever they stand on the frontline of the struggle for Indigenous rights and sovereignty.

Extracting Knowledge

The concept of Relationality as a central concept in Indigenous ontologies has been articulated by several Indigenous scholars, such as Nick Estes, Winona LaDuke, Lauren Rynan, Edward Valandra, and Melanie Yazzie, and non-Indigenous ones (Donna Haraway, Ana Tsing). It posits that the interactions between humans and other-than-human beings are within a lateral scale model where they coexist side by side, rather than privileging a vertical hierarchy. In many Indigenous societies, the land, water, humans, animals, and plants are all interdependent and relate to one another as kin. I believe it is important here to highlight the oeuvre of Caribbean writer and cultural theorist Edouard Glissant who published *Poétique de la Relation* in 1990. A foundational book in the theory of relation, it envisioned the future of global geopolitics (geopoetics) as orchestrated by *creolization*, a process repositioning cultures and histories within intertwined networks of political, social, and aesthetic relations, resting on improvisational dynamics which are key to their creative renewal. Glissant theorized that in creolized cultures, the right to opacity and to withhold transparency were forms of resistance against the dominant structures (see chapter 8 in this volume for a discussion of epistemic untranslatability as critical to the fight against the structural violence and dispossession in the Esmeraldas).

Relationality is sometimes used as an antonym to the extractivist ideology and pits Western profit-led extractivist societies against non-extractive Indigenous societies. This critical tool to examine the ontological relations between various actors coexisting in Indigenous communities has been used at times in cultural and environmental discourses to support a misguided romantic vision of how Indigenous knowledge systems perform. As Nick Estes argues, the concept of relationality needs to be anchored in a living reality, uniquely experienced in various contexts, historically determined, and not viewed as an overarching configuration:

> This is a huge word in Indigenous studies right now — relationality — that I think has become mystified. We're not the Na'vi of *Avatar* running around plugging our brains into trees trying to download data. I go back to the buffalo, because buffalo relations really represent the form of relationality that we had with animals. It wasn't just this mystical kind of thing where we were communing with them outside of history. They represented a source of life for us in the sense that without the buffalo, we wouldn't be the Lakota people, by mere fact that we wouldn't have a food source. (Serpe, Estes, 2019)

A truncated and ultimately false interpretation of Indigenous ontologies fuels some environmental positions that look at Indigenous ontologies as extractable resources, fragments of which become part of discussions about the environmental crisis and integrated into institutional projects. Often contrasted with Western "scientific" knowledge systems, this process is found in some strategic initiatives in governmental and international agencies.[4] Indigenous knowledge viewed as a trope to be exploited in the interest of global recovery from an environment crisis largely shaped by the extractive industries, conceives Indigenous knowledge as a commodity: colonization is also a cultural enterprise. This position participates in a profound misunderstanding of the far reaching macroscosmic vision of the world encompassed in Indigenous ontologies and powering resistance movements. Indigenous ontologies cannot be part of another appropriation process that is oblivious to the epistemological frameworks in which these ontologies exist and are rooted.

In sum, the success of extractivist ideologies depends on a logic of exploitation, violence and subjectification to establish relations of domination, erasing anything that could hinder the global expansion of capitalist and neoliberal systems, using armed violence against resisting Indigenous populations, peasants, and Black communities in the Americas, a violence that the #NoDapl movement made visible globally.

Organization of the Book

Digging Earth: Extractivism and Resistance on the Indigenous Lands of the Americas is organized in three thematic parts, each bringing a different, complementary dimension to the various Indigenous resistance movements against resource and knowledge extractivism. In the first part, Mining Indigenous lands and Knowledge, essays and artists' portfolios focus on extractivist processes. The volume opens with Shinnecock photographer Jeremy Dennis' work and an essay by Erin Joyce, both about the exploitation and erasure of Indigenous histories and cultures on Long Island. Dennis's "On This Site: Indigenous Long Island," is a photography project that traces the origins of culturally and historically significant places on Paumanok (Long Island) and records the Indigenous knowledge handed down through oral histories and archeological research. Paumanok is the homeland of the Unkechaug, Setauket, Shinnecock, Corchaug, and Montaukett peoples and Dennis's photographs provide powerful testimonies of their presence over thousands of years. In her essay, Erin Joyce demonstrates how Dennis documents the Shinnecock resistance against the appropriation of their homeland, objects, and sacred sites since 1640 while it examines the role played by museums and cultural institutions in the colonial extraction of Indigenous knowledge.

The extraction of minerals on Diné and San Carlos Apache homelands is the subject of "Mining Indigenous Land: Decisions and Opinions. Uranium and Copper in the American West." Susanne Berthier-Folgar addresses the notion of "sacrifice zones" presented by the US federal government as inevitable collateral damage from mining to support the twentieth and twenty-first centuries' industrial and nuclear development in the United States. The chapter underlines the consequences of past uranium mining on Diné Bikéyah and studies the copper mining project on the sacred site Chi'chil Biłdagoteel (Oak Flat) of the San Carlos Apache homeland. Organizations such as Apache Stronghold, the Black Mesa Water Coalition and To Nizhoni Ani "Sacred Water Speaks" continue to challenge the agreements between the federal government and the mining industries.

The relation between mineral extraction, the history of settler colonialism in Chiapas, and the resistance movements in the region, are the themes of Lynn Holland's "Mining in Mexico and the Land Defenders in Chiapas." Land Defenders, the People's Front for the Defense of Soconusco, and the Zapatistas offer alternatives to the extractive-exploitative model in the region. Holland underlines how the consolidation of the nation-state and modernization theory contributed to the implementation of an extractive economy and strategies such as the partition of the lands, the displacement of Indigenous communities and violent actions against communities who stood in the way.

The protection of water organized by Water Defenders in several struggles throughout the Americas is a recurring theme in this volume. The artist Carolina Caycedo works with riverine communities in Colombia. Using handmade fishing nets, atarrayas, Caycedo creates large hanging sculptures, *Cosmotarrayas*, that weave visual narratives emphasizing the relation between these communities and the Yuma river. In the accompanying essay, "The River as a Common Good," Carolina Caycedo and Jeffrey De Blois demonstrate how Caycedo's artistic practices and interventions expose the destructive practices of the hydroelectric companies, while highlighting the cultural and ontological role of the various bodies of water in the region. Caycedo's work recognizes water as a living entity, a topic addressed in Valandra's chapter, "Mni Wiconi, Water Is Water [is more than] Life," and in Carcelen-Estrada's and Sault's contributions.

Indigenous voices in resistance movements rising against the exploitation of the lands and waters resound loud enough to echo throughout the American continent. The second part of the volume: Resisting Extractivism: the Centrality of Indigenous Voices, affirms the defining role of women activists, grassroots organizers, and Water Defenders. Edward Valandra, in "Mni

Wiconi: Water Is [more than] Life" positions the #NoDAPL movement led by the Oceti Sakowin Oyate within the larger context of histories of Indigenous resistance and underscores the fundamental claim of Indigenous sovereignty in the struggle against the implementation of the Dakota Access Pipeline (DAPL). The Water Protectors at the Oceti Sakowin camp view Mni Wiconi, the Missouri river, as a relative, not a quantifiable commodity. Mni Wiconi is alive, Valandra argues, and water must be treated as kin: *we have to be a good relative*. The solidarity between the many Indigenous nations and their representatives, non-Indigenous people, and the plurality of nationalities, gender, age, and sexual identities of the participants made the Indigenous resistance at Standing Rock a worldwide movement that brought attention to the global struggles of Indigenous communities.

In "Native Feminisms and Contemporary Indigenous Art: Emma Robbins and Angelica Trimble-Yanu Counter the Gendered Violence of Extractivism," Elizabeth Hawley highlights the work of two Indigenous artists who rise up against the neocolonial gendered violence of extractivism on Diné homeland, Makȟóšiča (Badlands), and Ȟe Sápa (Black Hills) of South Dakota. Robbins's and Trimble-Yanu's feminist practices and artworks are mapped out by the ancestral connections existing between women kin and the earth in their homelands. Hawley acknowledges her position as an author standing with the artists and practicing an "allied scholarship" needed to deconstruct Western feminist narratives and art history canons.

"Water, Women and Resilience in the Andes of Peru," draws attention to grassroots movements led by rural and Indigenous women in the arid coastal regions and highlands of Peru. Nicole Sault carefully establishes the connection between the threats to water on Quechua and Aymera homelands and mining corporations. The construction of large dams further depletes the water from agricultural communities and results in environmental toxicity and population displacement. The author stresses the solidarity practiced by several Andean communities who organize a common resistance to defend the rights of water.

Antonia Carcelen Estrada's "Extractivism in the Country of Good Living: The Dystopian Reality of Black and Indigenous Struggles" establishes the central role of women in the resistance movements to extractivism in the Indigenous Amazon and the Esmeraldas communities on the Pacific coast of Ecuador and explains how the militarization of the conflicts applies continued violence against women in both regions. The author highlights the importance of political and cultural solidarities in plurinational and intercultural alliances that bring together several Andean and Amazonian communities (Kichwas and Shuar) in the Pachakutik movement. In the Black Pacific, where women of African descent in the Esmeraldas build their own transnational resistance funded on African epistemic practices, Carcelen-Estrada explains how untranslatability is a tool resisting the appropriation of knowledge as an "ornament to development discourses".

In the third part of the volume: Taking Back Indigenous Lands, Waters and Epistemes, authors and artists present on-the-ground action and theoretical propositions that generate alternative models of relation and interconnectivity based on Indigenous knowledge systems. Through the lens of photographer Will Wilson in "Connecting the Dots for a Just Transition," we see the abandoned uranium mines and disposal sites on Diné Bikéyah. Wilson claims the urgent need to heal these sites and surrounding lands and the fundamental importance of Diné knowledge systems in accomplishing it. Leetso, "yellow dirt," is a substance linked to the underworld that cannot be disturbed, and in response to its destructive power, the Diné choose corn pollen, a yellow substance that brings life. In Wilson's view, similar to that of other contributors in the volume, the natural environment is interwoven with the life of humans and other beings; kinship is articulated through historical, religious, and cultural narratives and is evidenced in Kaila Schedeen's essay, "Will Wilson: Connecting the Dots of Uranium Extraction on the Navajo Nation."

Kelsey Leonard's "Regenerative Relationships of Harvesting Clay for Healthy Soils and Communities in the Shinnecock Nation," discusses clay as an agent of social and cultural construction and reclaims an ecological model deeply embedded in the historical narratives of Indigenous Peoples of the Atlantic shores, including those on Paumanok (Long Island). On the Shinnecock homeland, the harvesting and usages of clay and ceramics relate to a deep experience and knowledge of the soil and Indigenous techniques. Leonard advocates rematriation in clay and ceramics practices —evidenced by the work of artist Courtney Leonard, the Indigenous women's practice of restoring ancestral relations between Indigenous Peoples and their homeland, a fundamentally anticolonial and anti-extractivist practice.

In "Ittoqqormiit Through the Pinhole: Exscribing the Ethno-Aesthetics of Pia Arke," Morten Søndergaard takes the work of late Inuit artist Pia Arke through a process of decryption that focuses on the idea of spatial displacement and ethnoaesthetic transformation. The author examines Arke's photographic practice as it encrypts notions of in-betweenness and belonging. The process informs the daunting photographs of the small towns of Ittoqqortoormiit, where she was born, and Nuugarsuuk. Her photographs claim the "reremembering" of fragments—places, memories, peoples, and exscribe an archive of untold, forgotten stories, where a space of "contact zones" emerges. Extractivism takes a personal resonance as she examines both her own history and the colonial extraction affecting Indigenous lands.

The collective DesertArtLAB works with the desert environment where the artists grew up and live. April Bojorquez and Matt Garcia use Indigenous plants—nopal, yucca, agave, in eco installations, community-based projects and exhibitions, and claim abandoned or misused spaces in Arizona and other parts of the Southwest. In her essay, "DesertArtLAB: Ecologies of Resistance," Rachel Zimmerman writes about DesertArtLAB's installations and on-the-ground community engagement tactics and how they provide nourishment and knowledge rooted in Indigenous ecological models. With *Field Site* and the *Desertification Cookbook* the artists affirm the long-standing relationship of Indigenous Peoples with desert plants and offer a healing adaptation to the climate urgency.

Juan Pablo Pacheco Berajano examines the relation between water and technology in "Humid Telepathy," and unravels our perception of communication technologies as invisible. Transoceanic routes were established with the first European expeditions, the beginning of exploitative capitalism. Today an underwater flow of data organizes the circulation of information capitalism through cable systems and data centers using extracted minerals for their microprocessors and technical apparatus. The author contrasts this process with that of telepathic perspectivism, a relational form of communication in Amerindian environments, mediated through water, plants, minerals, sacred architectures, and human communities. Pacheco Berajano reveals the presence of histories and knowledges contained in bodies of water. Unmediated by extractivist technologies, humid telepathy is a communication network, unfollowing linear structures and moving instead incessantly like water.

In conclusion, this volume cannot in any way be considered a complete account of the existing extractive technologies, of their impacts on Indigenous homelands, or of the Indigenous resistance movements. It is an attempt to stand and speak with those who voice, with urgency, the necessity of working in solidarity to combat the scenarios created by colonial and neo colonial extractivist politics leading to the destruction of the earth. The artists, activists and thinkers in *Digging Earth: Extractivism and Resistance on Indigenous Lands of the Americas* bring forward new networks of relation and new visions of being with the world. This is an invitation to hear the sounds of the lands and the rivers, rumbling and muttering around us.

Notes

[1] The Artic region, considered a part of the Global North, include Indigenous communities who are affected by extractivism, in Canada, Greenland, and Alaska in the US. Different groups comprising the Inuit, Yupik and Inupiat are subjected to oil drilling and gas extraction with dramatic impacts on their lands, fauna and flora. The Biden administration for example signed an agreement to open the eight billion Willow oil project led by ConocoPhillps in Alaska in March 2023. See also: https://storymaps.arcgis.com/stories/8384836f8a4e4aa49308fa5afb0d319b

[2] On August 8, 2023, President Biden signed a proclamation establishing a new national monument, north and south of Grand Canyon National Park to protect the area against uranium mining, including in the Pinyon Plain Mine. It also protects sacred sites to the Havasupai, Hopi, Hulapai, Paiute,

Diné, Yavapai-Apache, Zuni and Colorado River Indian Tribes from new uranium mining projects.

[3] The General Allotment Act, also called Dawes Severalty Act (February 8, 1887), is a US law providing for the distribution of Indian reservation land among individual Indigenous Americans and the fostering of individual property. The result of the Allotment Act was to break tribal social structures. In addition, a provision of the Act allowed any "surplus" land to be made available for public sale to ranchers or settlers. By 1931 the white settlers had acquired two thirds of the 1887 Native territory.

[4] This process is found for example, in some initiatives of international agencies such as the UN Environment Programme (UN Environment Programme, 2021), the UN System Chief Executives Board for Coordination (UNSCEB, 2020), and the Food and Agricultural Organization (FAO). The Center for Sustainable Development, an international private institution that proposes programs adapted to training NGOs, teaches participants how to use aspects of Indigenous knowledge in the implementation of solutions to the climate crisis. See for example: https://www.unep.org/news-and-stories/story/how-indigenous-knowledge-can-help-preventenvironmental-crises; https://unsceb.org/topics/indigenous-peoples; https://www.fao.org/indigenous-peoples/indigenous-peoples-food-systems-book/en/

References

Acosta, A. (2013). Extractivism and neo-extractivism: two sides of the same curse. In Lang, M. and Mokrani D. (Eds.), *Beyond Development: Alternative Visions from Latin America*, pp.61-86. Quito: Transnational Institute, Fundación Rosa Luxemburg.

Dhillon, J. (2019). Notes on Becoming a Comrade: Indigenous Women, Leadership and Movement(s) for Decolonization. *American Indian Culture and Research Journal*, 43(3), pp. 41-54.

Dhillon, J. (2021). Indigenous resistance, planetary dystopia, and the politics of environmental justice, *Globalizations*,18(6), pp. 898-911. 4 DOI: 10.1080/14747731.2020.1866390

Estes, N. (2019). *Our history is the Future*. London, Brooklyn NY: Verso.

Grand Canyon Trust (2022). *Map of Active Mining Claims Within Grand Canyon Withdrawal Area, May 2022*. Available at: https://www.grandcanyon trust.org/map-active-mining-claims-within-grand-canyon-withdrawal-area-may-2022

Lepenies, P.H. (2013). *Art, Politics, and Development*, Philadelphia, Temple University Press.

Martin, M. (2017). Latin America Unites Against Hydro Dams, Environmental Destruction, *Grassroots International*, October 25.

Newland, B. (2022). *Federal Indian Boarding School Initiative Investigative Report*, May. Available at: https://www.bia.gov/sites/default/files/dup/inline-files/bsi_investigative_report_may_2022_508.pdf

Serpe, N., Estes, N. (2019). Indigenous Resistance is post-Apocalyptic, *Dissent*, July 31. Available at: https://www.dissentmagazine.org/online_articles /booked-indigenous-resistance-is-post- apocalyptic-with-nick-estes

Stern, J. (2021). Pipeline of Violence: The Oil Industry and Missing and Murdered Indigenous Women, *Immigration and Human Rights Law Review*, May 28. Available at: https://lawblogs.uc.edu/ihrlr/2021/05/28/pipeline-of-violence-the-oil-industry-and- missing-and-murdered-indigenous-women/

Urban Indian Health Institute, (2018). *Missing and Murdered Indigenous Women and Girls,* Report, November 14, Available at: https://www.uihi.org/wp-content/uploads/2018/11/Missing-and-Murdered-Indigenous-Women-and-Girls-Report.pdf

Yazzie, M. (2018). Decolonizing Development in Diné Bikeyah, *Environment and Society,* 9, pp. 25-39. Available at: https://www.jstor.org/stable/10.2307/26879576

Part 1

Mining Indigenous Lands and Knowledge

Chapter 1
Extractivism

Erin Joyce

Taking it by force. Remove it by grabbing it and keeping it for yourself. Keeping it and holding onto it violently, without yielding or bending or offering concessions. Extracting its energy, its resource, its lifeways and using it. For Jeremy Dennis (Shinnecock, b. 1990), extraction and extractive entities are part of the lived reality for Shinnecock peoples living in their ancestral homelands on Long Island, just outside of Southampton, NY. The 1,000-acre reservation is home to half of the Shinnecock population, which has a current enrollment of 1,589 members. "I always tell people, it's no different from anywhere else, but the reservation is almost entirely residential structures and everyone you know is your cousin essentially," said Dennis about his homelands. The photographer and community organizer explores various and nuanced facets of Indigenous life, from identity and representation, actualities of settler colonialism, all through a direct form of documentation. Dennis' work aggregates the archaeological, anthropological, the historical, and the oral and distills it into questions regarding his Shinnecock ancestry and culture. Questions of "where did my ancestors live? Why did they choose these places? What happened to them over time? Do these places still exist?" Through the exploration of these questions, Dennis uncovers the inextricable connection between place and memory. Perhaps a notion of a place acting as a container for memory – memory of time, a people, of violence(s), and current battles being fought.

Dennis, who lives and works in Southampton, New York on the Shinnecock Indian Reservation, has seen first-hand the impact that contemporary land grabs have had on his community. "Nationwide, lands have been stolen from Indigenous peoples to various degrees. Here in what is known as The Hamptons, we clash with Southampton often over our blatantly stolen Shinnecock Hills. Because the real estate industry is so large and tax-generating, we rarely see eye to eye with Southampton town," said Dennis. European settlers first arrived on Shinnecock in the 1640s and began a sustained occupation and extraction of lands – seizing territory from the Shinnecock people – removing them from these ancestral lands, exploiting their labors, and disappearing them from settler-colonial view. This coloniality and privileged mentality is by far not a thing of the past. "On the reservation, there are great youth and elder programs to take care of those groups and foster community," said Dennis. "I think between those two age groups, however, there are fractured groups from politics and family divisions. I try to use the arts whenever I can to bring people together with this in mind." Dennis' photographic practice surveys the textured topography of identity, land, and cultural practice. His work points to a complex duality of living on a sovereign Indian reservation within the strictures of a colonized nation, amplifying the struggles Shinnecock people face. Due to the increasingly titular precarity of racial divisions and tensions in the United States, it has become increasingly important to the artist, and community organizer, to offer complex and compelling representations of Indigenous life. "In the past I have done work around more human experiences and philosophy," he shared. "But there is such an urgency in my work around Shinnecock identity [currently.]"

Since becoming a Tribe recognized by the Federal Government in 2010, some improvements have been made in the form of aid towards the Shinnecock community – in specificity to self-sufficiency of the Tribe and economic development. These are most visible in the way of relief from historic storms, which have increased in frequency due to climate change and environmental collapse, as well as the devastating effects the Covid-19 pandemic had on Indigenous communities in North America, both of which held and hold great amounts of urgency to them. "We will potentially see climate change raise the sea level in the next couple of generations," said Dennis. "Our nation is at sea level. It is urgent to regain our exploited and extracted stolen lands to allow us to remain on our aboriginal territory uphill in the Shinnecock Hills."

Battles over the land, who owns it, and who has a right to it, have embroiled the Shinnecock people with their Southampton neighbors. One such example is the quest for Tribal gaming. For the past two-decades the Tribal leaders of the Shinnecock community have fought to build a casino outside of Manhattan to create much-needed sources of revenue to lift the Tribe from poverty and bolster social and community programs on the reservation. This fight has been unsuccessful (Kilgannon, 2021). "I believe it [gaming] will help through the revenue it brings in, assisting in paying for Tribal jobs, road maintenance, and more, but gambling addition is a serious issue that affects casinos and online gaming," shared Dennis. In more recent years, the Tribe has pivoted to other manifestations of gaming, some of which have been proposed to take a physical form on their reservation – a proposal that the ultra-elite residents of Southampton have been ferociously protesting. The Hamptons, known for white sand beaches, quaint village shops and restaurants, and ultra-luxurious mansions, that sit empty the majority of the year, is the playground of wealthy New Yorkers (and others) in the summer months. These same individuals are in opposition to the development of a casino by the Shinnecock Nation, citing that it could bring crime and would mar the pristine landscape – a hypocritical stance for a population of people who have scarred the landscape with monstrously overbuilt estates. "We at Shinnecock value the land above everything else, and so extraction in our region is seeing our neighbors exploit the land and develop it so outrageously," said Dennis. "We see our identity and the land as being one. Southampton and its residents extract wealth and affluence from stolen lands." In recent years, Dennis has spent time evaluating what you see of Shinnecock people culturally within Southampton, noting that nothing in town references the fact that Shinnecock people continue to inhabit these lands, not even five minutes down the road. "We have remained here for more than 10,000 years," he said. "Aside from the caricatures [of us] on the sides of garbage pails in town, [we're not seen there.]" Dennis recounts the blithe ignorance of many Americans to the histories of Indigenous peoples, but also acknowledges a willingness to remain so – in the Hamptons, all the way up to the New York State Governor's office, "we observe a doubling down of ignorance and profit [over people.]" Though this extraction is not in the literal vein of extraction one may think of, that of mineral deposits, uranium, oil, or coal mining – there are very real consequences and aftermaths that occur. "I strive to continue my ancestors' tradition of storytelling and showcase the sanctity of our land, elevating its worth beyond a prize for the highest bidder," he shared. "One daily issue I have relates to land use and how Southampton is largely dotted with vacant housing and storefronts, and how this relates to the traffic every single day in our area. People are driving hours from out of town to service homes and businesses that they themselves may or should be able to live in. It is all backward and in service of the wealthy, which Southampton town seems to be happy to serve."

In his series, *On This Site – Indigenous Long Island* (2016-present), Dennis addresses and redresses colonial occupation of Indigenous lands in so-called North America and brings into view what has been hidden in plain sight – Indigenous existence – making the invisible visible. Through the study of archaeological and anthropological records, oral stories, and newspaper archives, Dennis captures photographic representations that illuminate the 10,000-plus year presence of Native peoples on Long Island. By excavating through the sediment of Shinnecock past, Dennis unearths the issues that have historically and presently plague Indigenous communities. Histories of land seizure and dispossession, settler-colonial attempts and forced assimilation of Native bodies, and campaigns of ethnocide and genocide have left a residue that has obfuscated the resiliency and survivance of Indigenous peoples of the Western Hemisphere, Indigenous lifeways, environmental stewardship, and the constant struggle to uphold tribal sovereignty (Wizenor, 2000). There is a cinematic quality of the images, a quality pushes its weight against the carcass of filmic misrepresentation of Indigeneity. The "celluloid Indian" as coined by Jacquelyn Kilpatrick in her 2006 text of the same name, has represented and propagated innumerably damaging amounts of imagery of Indigenous peoples in the Americas – creating reductive and problematic readings of what it means, looks, and sounds like to be Native (Kilpatrick, 2006). "Nowhere have indigenous peoples been more poorly misrepresented than in American movies," said Dennis. "My images question and disrupt the post-colonial narrative that dominates in film and media and results in damaging stereotypes, such as the "noble savage" depictions in Disney's Pocahontas." The photographs give brief, eerily intimate glimpses at significant Shinnecock cultural sites – showing the paradox between the colonial atrocities these places, and the people who inhabited and still inhabit them have undergone, and beautiful resilience that was and is necessary for these sites to endure despite the theft and conflict with the town of Southampton. The works create a portal into an environment that, though visually dynamic and soaked in aestheticism, create conversation surrounding these uncomfortable aspects of colonial occupation. "Despite four hundred years of colonization, we remain anchored to our land by our ancient stories," shared Dennis.

Beyond the very real physicality of extraction of mineral resources and seizure of territory(ies) in some Indigenous communities, and the dubious annexation for development of land in others, is this need to consider how Indigenous culture, heritage, and visual history is extracted and dispossessed of the people. "Some museums continue to have our sensitive objects, which is a form of extraction," he shared. For example, when we consider the ways in which visual and popular culture shape our perceptional reality of the world, we must also question the role museums play within that landscape. What are the mechanisms that enable museums, specifically art and anthropologically aligned museums, to act as perpetrators or accomplices of structurally violent systems in the United States and Europe both historically, from the colonization of the western hemisphere to this current moment? Are they merely benign institutions that present and (re)present the cultural matrimony and patrimony of a community or are there more serious consequences ensconced within the skeleton of what a museum structurally is and how it functions. If we include museums in the same vein of filmic misrepresentation and captured depictions of Indigenous peoples perpetuated by the media, then we see can see them as contributors to harm. Art and museums are not typically viewed as foundational elements of society in terms of survival or prosperity; however, they form a structural element that enables and shapes systemic racism and violence via the prism of the museum through the inclusion or exclusion of marginalized communities and voices.

If we consider the ways in which cultural institutions and museums operate as agents of media — formalized spaces that house and disseminate what is thought of as an official discourse, message, and meaning — we can begin to unpack and position these entities not only as places that present and (re)present culture in a space of purported neutrality, but as institutions that build meaning constitutively around the objects and cultures they display; often without consent, context, or concern from or for the communities they have extracted from, often across multiple registers of meaning. Museums have a self-bestowed power to reconfigure past, present, and future, perpetuating a level of colonial amnesia via the lens of Euro-American myth-making regarding any individual or group that can be deemed 'other.' Much of this is resultant due to internal policies on collecting and exhibiting works, institutional power structures, and even wage gaps that are seen played out in museums – with large salaries paid to executive teams and meager wages paid to mid-level staff and frontline workers; the need to look at these armatures of operation and how they perpetuate erasure of Native peoples both historically and in the future. There is a need to examine the various modalities employed by museums that perpetuate violence and erasure of non-white artists, communities, and cultures, through exhibition making and exhibition policies, collections policies, hiring policies, board recruitment, funding models, and the ways in which museums create or restrict access. "Because we have 10,000 years of ancestors in these lands, our cultural resources are virtually everywhere you dig," said Dennis. "But many of these sacred objects have been extracted for personal collections as well, if not discarded, destroyed, or sold." We must consider the ways in which Indigenous communities in North America have been (re)presented/mis(re)presented, and had their cultural material extracted from their homelands. Again, in framing 'media' as a stand in for 'museums' and viewing museums as an entity that functions as a technology – understanding that a technology at its core is created as a means to an end, then the museum as technology can be explored as an entity a means to disseminating an official discourse, meaning, or truth. With that, we have to come to terms with the reality that truth is constructed and perceptional, and that museums enact culture and perpetuate its inherent biases whilst also capitalizing on the potential of change. "As a practicing artist, you are often tokenized in art institutions and galleries as an exotic 'other,'" said Dennis, noting the issues surrounding representation of contemporary Indigenous artists in museums, as well as noting a reckoning regarding museum ownership of historic Indigenous visual culture. "Historic objects are typically problematic because they were obtained out of desperation from the person giving them away or due to conquest against Indigenous peoples and the spoils of war and genocide." Dennis notes that Indigenous objects, more often than not, should not reside within the colonial confines of a dominant culture/non-Native museum. "Some of these objects can be repatriated into the ground, others can be returned to Tribal collections and museums."

Again, applying media as moniker for museums and cultural institutions overlays the idea that what is encountered in a museum, either in exhibition format, programmatically, or digitally including websites, webinars, and social media, shapes an unspoken understanding that the viewer is experiencing an accurate representation of what is being discussed or presented. Within the context and reference to Indigeneity in the Western Hemisphere – there have been and will continue to be very real consequences that harm Native communities, psyches, and cultural heritages. "There needs to be more acknowledgment and education around Native people, history, and contemporary issues," shared Dennis. "I think this is number one because generally the public starts at zero and simple understandings of how we have gotten to where we are today would alleviate much of the racism and stereotypes that exist today." If the public does have a semblance of knowledge(s) around Native people, it usually is derived from filmic and museum representations of them, and not through first person voices from the communities and elucidation of colonial atrocities. "I had the realization in my mid-20s that learning about my ancestry was important and empowering, and wondered why my education did not include this local history of Shinnecock people. I started with *On This Site* in 2016 to fill this gap."

The on-going photographic series, as mentioned above, brings a level of visibility – or perhaps not just visibility – but context to the landscapes of Shinnecock, and tangentially – to the Shinnecock people. There lies within the work, a reclamation of agency and authority of representation – even down to the medium of choice – the photograph. "Photography is a powerful tool to produce evidence of something existing," stated Dennis. "But historically it was used negatively towards Native people, telling the story of our disappearance and the need to document us before that

happens." remarking on its instrumentalization by photographers like
Edward Curtis to create fixed, inaccurate views of Nativeness and state
sanctioned attempts at extinction of Indigenous people. "Today, I use it to
show our strength and resilience." That said, Dennis treads carefully with
the medium, acknowledging its power for good as well as for bad.
"Photography is always exploitative," he said. "I am always evaluating
who is taking the photo of whom, who will prosper from the photos taken,
who will have the rights to use and forbid the photo usage, and what story
can a single image tell or not tell."

The work is an action. A means to communicate, protect, and illustrate the
very real need of Shinnecock people to be seen, heard, and listened to.
When asked what reparations towards his community would look like,
Dennis replied, "First our lands being returned and allowing Shinnecock a
seat at the table when it comes to land usage as well as monetary
compensation for the generations of stolen land exploration, and our
ancestors being sent away as slaves and exploited for their labor locally. I
hope one day my work will have no meaning because our neighbors and
the United States will have taken action to undo their problematic and
colonial ways."

References

All quotes in the text are: (unless otherwise specified)

Dennis, Jeremy. February 2, 2023, zoom conversation.

Dennis, Jeremy. February 20, 2023, zoom conversation.

Dennis, Jeremy. March 20, 2023, email communication

Kilgannon, C. (2021). Why the Shinnecock Tribe Is Clashing with the Hamptons' Elite. *The New York Times*, April 22, 2021. Available at: https://www.nytimes.com/2021/04/22/nyregion/casino-hamptons-shinnecock.html.

Kilpatrick, J. (1999) *Celluloid Indians: Native Americans and Film*. Vancouver: Lincoln, University of Nebraska Press.

Vizenor, G. R. (2000). *Fugitive Poses: Native American Indian Scenes of Absence and Presence*. Lincoln: University of Nebraska Press.

On This Site - Indigenous Long Island: Jeremy Dennis

Figure 1.1
Niamuck I Canoe Place I Shinnecock Canal, 2017
Archival Inkjet Print, 30 x 40 inches

Niamuck was once the primary location of settlement for the Shinnecock people prior to the current Shinnecock Reservation. From the current Shinnecock Canal to the southernmost land tip, the Shinnecock people existed before contact. Shinnecock people primarily resided in the area until ca. 1703, though historical maps show continued presence until the mid-19th century. In 1791, Rev. Paul Cuffee organized a Congressional church in the Niamuck area. Part of the church remains in the area while the other half was moved to the Shinnecock Reservation and is still used today. Cuffee was buried on the spot where the church once stood. This place is now known as the Shinnecock Canal.

Figure 1.2
Circassian Shipwreck Site, 2019
Archival Inkjet Print, 30 x 40 inches

On December 30th, 1876, freight ship Circassian wrecked off the coast of Mecox Bay in Bridgehampton. Ten members of the Shinnecock tribe were among those who drowned while attempting to rescue to ship's crew and cargo.

Since the tragedy, the Shinnecock Nation has held remembrance gatherings that include drumming, dance, and feasting to remember ancestors who gave their lives to support their community.

The Heroes of 187
David W. Bunn, James Franklin Bunn, Russel Bunn, William Cuffee, George Cuffee, Warren Cuffee, Oliver Kellis, Robert Lee, John Walker, Lewis Walker

Figure 1.3
Fort Shinnecock, 2017
Archival Inkjet Print, 30 x 40 inches

This sacred glacial erratic marks the location of what may have been both the Shinnecock Fort and June Meeting location in the Shinnecock Hills. There have been many references to a contact-period Shinnecock fort, but the specific location has likely been disrupted by development. June Meeting is a Presbyterian and Algonquian inspired celebration and gathering for Eastern Long Island tribes started by Reverend John Cuffee in the 1700s and continued annually on the first Sunday in June. It's a time of dance, feasts and the passing down of stories and traditions. The Unkechaug tribe continue this traditional seasonal celebration in the western town of Mastic. According to Shinnecock oral history, this site, similar to other council rocks, were the places for indigenous leaders to gather for important meetings. Today, this land is located off of the current bounds of the Shinnecock reservation. The town of Southampton bought and preserved the area using its Community Preservation Fund for its cultural significance.

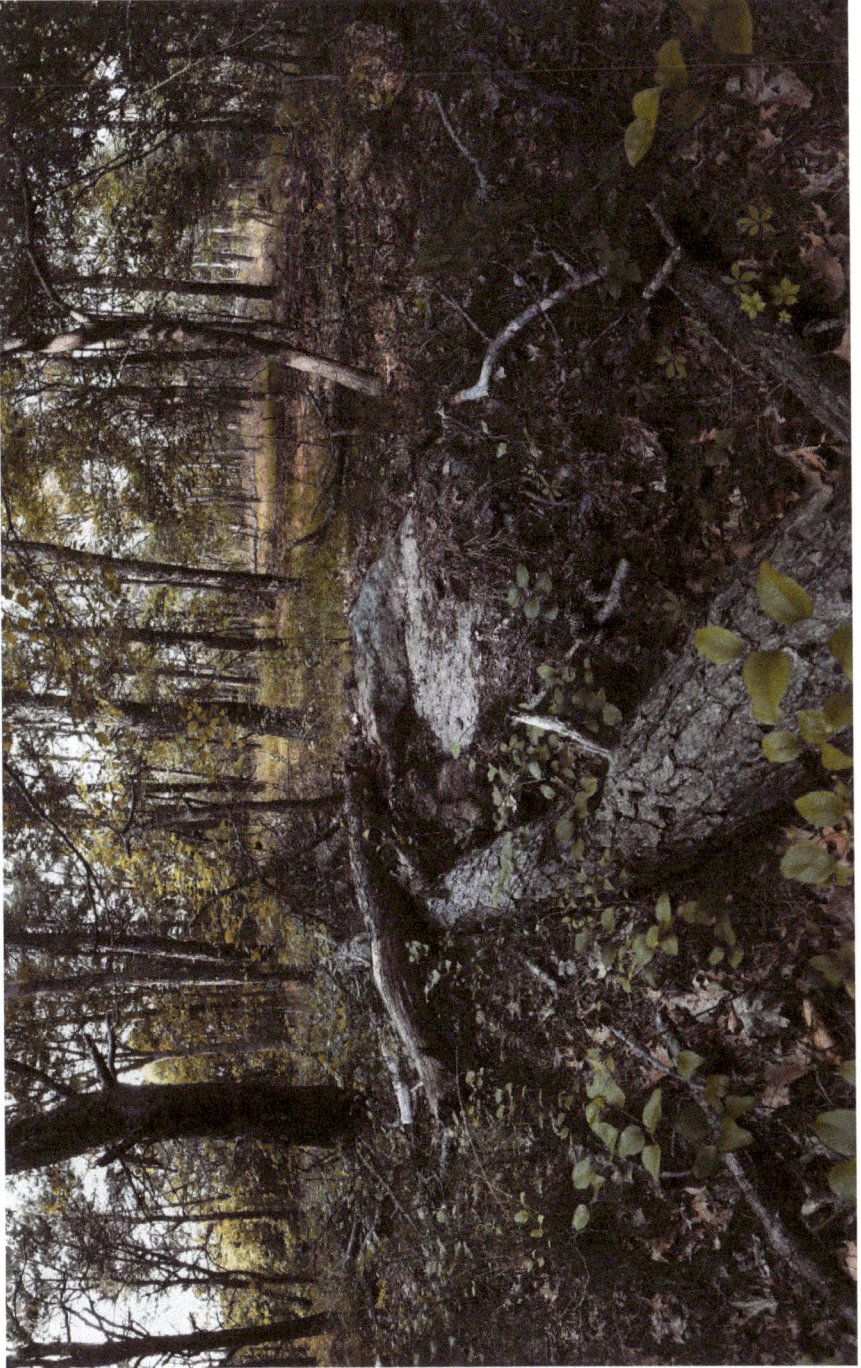

Figure 1.4
Hawthorne Site, 2018
Archival Inkjet Print, 30 x 40 inches

On Monday, August 13th, 2018, skeleton remains were found during residential development on Hawthrone Road in the Shinnecock Hills. The developers and homeowners contacted the Southampton Town and Suffolk County police department, who quickly disturbed the ground further for evidence of recent criminal activity. Along with human remains, a glass bottle from the 17th-century contact period was found, indicating a likelihood of the remains being of Native American descent with burial offerings. The Shinnecock Indian Nation arrived on the site soon after the detectives with the goal of overlooking the development. If the remains are from Native descent, the tribe encourages the town to use its Community Preservation Fund to preserve the lot and respect the burial.

Figure 1.5
Sugar Loaf Hill, 2021
Archival Inkjet Print, 30 x 40 inches

Sugar Loaf Hill is an Orient Period burial site facing southeastern, the only Orient burial site known outside of the North Fork of Long Island. During the 20th century, despite being known and marked on maps as early as 1797, the burial grounds were desecrated and developed for a contemporary residence. After years of pleas by the Shinnecock Indian Nation to save ancestral burial grounds in the Sugar Loaf Hill area of Southampton, there was a victory in 2021 after 30 years: The Southampton town board voted unanimously to green-light a $5.3 million purchase of a conservation easement for 4.5 acres at the peak of Sugar Loaf Hill with the assistance of the local non-profit Peconic Land Trust.

Chapter 2

Mining Indigenous Land – Decisions and Opinions: Uranium and Copper in the American West

Susanne Berthier-Foglar

Two case studies about minerals extraction on Indigenous land in the Southwest of the United States highlight the evolution that took place between the uranium extraction on Navajo land from the 1940s to the 1980s and the proposed Resolution Copper mine on Apache land. While uranium mining during the Cold War did not consider the well-being of Indigenous Peoples, who are still suffering from the impact of radioactive pollution of their land and bodies, the fate of the copper mine is still undecided. Legislation in favor of Indigenous beliefs—and related land-rights—has been passed, but the pressing need for raw materials might override the wishes of the Apache lobby to safeguard an area that was part of their homeland before Arizona's mining industry took over in the nineteenth century. Several issues are discussed, including: the resacralization of Indigenous land, the demand for "green" or "clean" energy without regard to its environmental and social cost, and the process of having to accept a sacrificial zone for the collective well-being of the population at large of the United States.

These two case studies, set seventy years apart, show how mining Indigenous land in the United States has evolved. The first case is about uranium mining on Navajo land from the 1940s to the 1980s, the consequences of which are still felt today. The second focuses on a copper mine in Arizona that has been planned since the mid-1990s and—as of 2023—is still in the pre-mining phase due to strong opposition by an Apache lobby. In both cases, those most affected are two Indigenous nations of the Southwest of the United States. The two minerals, uranium and copper, are, or have been, critical for the United States.

In 2022, the General Mining Act of 1872 became 150 years old. When it was passed, it was meant to promote the exploration and extraction of minerals, but it did not include any environmental provisions nor was there consideration for Indigenous rights or any form of human rights. It opened an era of extractivism that devastated western landscapes; they became sacrificial zones on which the white man's world was built without regard as to what was being destroyed.

This essay focuses on two of these sacrificial zones; one is recovering, painfully, and the other is in the making. In the temporal space between both, decades of activism, the implementation of new principles of operation by industry, and a fundamentally changed legislative environment regarding Indigenous Peoples have profoundly modified their agency.

Sacrificial Zone 1: The Navajo people, Uranium, and the Cold War

The search for uranium started after the United States entered World War II and shortly thereafter, the Manhattan Project was launched in Los Alamos, a remote area of northern New Mexico, to -before the Nazi regime- create the most powerful weapon the world had ever seen. As early as 1939, physicists Albert Einstein and Leo Szilard had informed President Roosevelt that Germany was working on a new weapon of unheard magnitude, but their advice was heeded only after the December 1941 attack on Pearl Harbor. The principle behind the new bomb was the fission of an atom of uranium that would release a chain reaction resulting in material being "spray[ed] outward like the shards of a grenade," bursting other atoms in the process (Zoellner, 2010, p. 3). Two atom bombs dropped on Japan ended the war and proved the superior destructive power of nuclear weapons.

When peace returned, the wartime partnership between the United States and the USSR fizzled out as the US's President Truman vowed to fight the spread of global communism (Amundson, 2004, p. 23). In 1949, when the world discovered that the Soviets were testing their own nuclear bomb, the US embarked on a program to build even more powerful weapons, ushering in an age of escalation and an increasing demand for uranium (Zoellner, 2010, p. 233).

Uranium often occurs with vanadium deposits, many of which exist in the West, and prospectors hoped to discover more. They met a Navajo shepherd, Paddy Martinez, who found the rocky outcrops of a major deposit near Grants, New Mexico, in the Four Corners area of Navajo Country (Amundson, 2004, pp. 23–24).

The Navajo consider their historic land to be defined by four cardinal mountains, anchoring the Diné—The People, as they call themselves—in their geographic surroundings. Mount Taylor, New Mexico, the Navajos' Tsoodzil, is their sacred mountain of the South, sitting on the massive reserves of the Grants uranium belt (Berthier-Foglar, 2011). Located between Albuquerque and Gallup, New Mexico, uranium country and Indigenous land intersect. The Navajo Nation and the Pueblos of Laguna and Acoma all have large uranium mines on their land, but the Navajo are the most impacted. Their iconic red mesas and spires, seen in John Ford's Westerns, where they were often used as a backdrop, represent an unrecognized and unmentioned danger zone.

When the mines and the mills started operating in the 1940s, the Navajo gladly accepted wage-work close to home, enabling them to supplement their failing pastoral economy. As in every uranium boomtown in the West, denial about the dangers of uranium ran strong and safety measures were largely nonexistent.

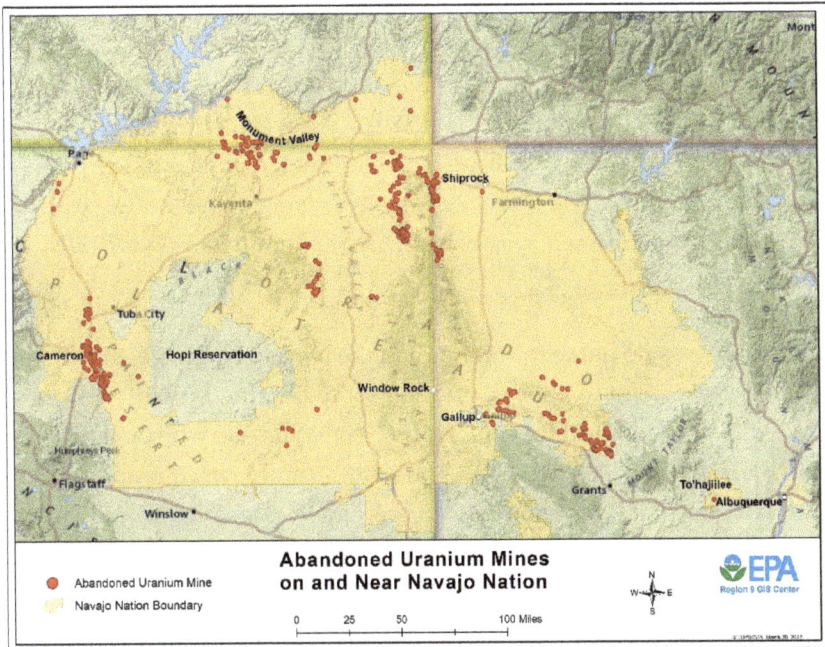

Figure 2.1. *Map of abandoned uranium mines on and near the Navajo Nation. EPA, 2016.*

No track record of mine openings was kept. Research done over half a century later shows six hundred abandoned uranium sites "either on or within one mile of Navajo land" but, depending on what is being called a mine, there may be up to one thousand sites (Hearing, 2007, pp. 21, 30; Brugge et al., 2006, p. 3). While the Jackpile mine was among the largest open-pit mines of the world, most mines were small diggings, dotting the landscape, some next to residential zones. Uranium mills were built near the larger mines to concentrate the ore, a process that involved crushing and grinding to obtain fine particles and chemical leaching to produce uranium oxide or "yellowcake."

After their shift, Navajo miners came home covered in yellow dust. Work clothes were shaken out or washed with the family's wash loads. Mine tailings were used in the concrete to build their homes. Navajo workers remember salvaging bits of equipment—a discarded mill grate to be used for cooking and drums for storage. Children would play in the tailings, roll around in the yellow dust, collect "marbles"—balls used in the milling process—and swim in the diggings next to their homes when rainwater had collected in a radioactive pit (Hearing, 2007, pp. 2, 84; Brugge, Benally, and Yazzie-Lewis, 2006, p. 4). Many Navajo homes did not (and still do not) have a municipal water supply and when the wells ran dry, men would bring back clean-looking water from the mines. Even in a dry country, mines require continuous dewatering, and using mine water seemed a good idea. Baby formula was made with contaminated water (Hearing, 2007, p. 118).

While there was no name for uranium in the Navajo language, they named its concentrate *leetso*—yellow dirt—and soon identified it with a monster, akin to the mythological monsters of their creation story (Brugge, Benally, and Yazzie-Lewis, 2006, pp. 2–3). The monster made itself manifest in the 1960s when many Navajo women became widows. The Navajo are cautious about discussing death and disease as they are related to ill omens and what Washington Matthews, an army major turned ethnographer called the "death taboo" in his early studies of Navajo culture (Matthews, 1994 p.78). The link to uranium was not immediately apparent as Navajo began to be afflicted by a variety of ailments, many of them deadly or debilitating. Health issues among miners and millworkers, including respiratory diseases; bone, skin and lung cancers; leukemia; kidney failure; and possibly birth defects (Brugge in Hearing, 2007, p. 40; Dawson and Madsen, 2011) only began to appear three to six years after exposure and, in the case of lung cancer, twenty years after (ACHRE, 1995). Moreover, the symptoms resembled those of other diseases. Worse, diseases affected not just the miners, but also their family members—the women and children who were not employed by the uranium industry (Brugge, 2000, p. 5; Brugge, Benally, and Yazzie-Lewis, 2006, pp. xvii–xviii; Eichstaedt, 1994, p. 55; Dawson and Madsen, 2011, p. 619; Arnold, 2014).

Government action was slow or nonexistent. A pilot study in 1950 pointed to possible health issues among the miners and was followed by a comprehensive study by the US Public Health Service, but the scientists were instructed not to warn the miners of the possible findings (Dawson and Madsen, 2011, p. 619). During the Cold War, turning "a blind eye and a deaf ear" seemed an acceptable option as disclosing information about the uranium production was deemed a national security risk (Pasternak, 2011, pp. 145–62; Brugge and Goble, 2002, p. 1415).

From the 1970s to the 1990s, the cancer death rate doubled on the reservation, while the US average declined slightly (Pasternak, 2006).

In 1979, two nuclear events happened three months apart in the United States. On March 28, 1979, there was a partial meltdown of a nuclear reactor at Three Mile Island, Pennsylvania, that led to the release of radioactive gases and iodine into the environment. On July 16, 1979, the dam of a tailing disposal pond at the Church Rock uranium mill, near Gallup, New Mexico, broke and released over 1,100 tons of uranium mining wastes mixed with 100 million gallons of acidic radioactive water. According to the Nuclear Regulatory Commission (NRC), the contaminated Rio Puerco "showed 7,000 times the safe standard of radioactivity for drinking water shortly after the breach was repaired" (Johannsen, 2013, p. 251). Church Rock was the United States' biggest uranium spill and it barely registered in the public eye while the Three Mile Island meltdown hogged the headlines (Johannsen, 2013, p. 260). The remediation at Church Rock is still ongoing, while the Three Mile Island cleanup was completed in 1993. Of course, population densities of both areas are widely different. However, Indigenous activists believe there is another reason. They came to consider that the slow remediation was deliberate, that it was part of what they saw as a nationwide "nuclear colonialism"–whereby indigenous land was to be sacrificed for the greater good.

More examples came to light.

It took over seventy years after the initial uranium pollution to have one thousand homes tested for radioactivity on the Navajo reservation (EPA & Navajo Nation Superfund, 2017). The extraordinarily long time it took to achieve this result reinforces the notion that the people affected were "the dispossessed and marginalized people who inhabit . . . zones of sacrifice" (Kuletz, 1997, p. xviii). By 2007, ninety Navajo homes still had dangerous levels of radioactivity but only two had been demolished and reconstructed (Hearing, 2007, pp. 124–28).

Cleanup has also been hampered by the numerous agencies involved and their diverse perimeters of action. The Nuclear Regulatory Commission regulates uranium recovery facilities but not uranium mining or abandoned uranium mine sites. The 1978 UMTRCA (Uranium Mill Tailings Radiation Control Act) mandates that the NRC take only mill tailings into account but not other types of radiation sources. The Department of Energy has authority over some sites, but needs a contract to intervene. Other agencies intervene with their specific agendas: the Office of Surface Mining, the Bureau of Indian Affairs, the Indian Health Service, the US EPA, and the Navajo Nation EPA.

When remediation was finally considered on a national level with the passage of the Radiation Exposure Compensation Act of 1990, it primarily targeted the white downwinders of the Southwest, those who lived in Nevada, downwind from the fallout of the atmospheric nuclear testing. It took ten more years to specifically include in it the miners and mill workers of eleven states, as well as an expanded list of diseases. Navajo miners could thus file for compensation. In 2022, the Radiation Exposure Compensation Act is still an incomplete mechanism for the Navajo as only pre-1971 exposure is considered, and the overwhelming paperwork requires professional help (Navajo Department of Health, 2022).

It seems that for the Navajo Nation, all the hurdles were in place to guarantee slow remediation, a situation that can be interpreted as "environmental racism" (Kuletz, 1997, p. xv).

The colonial practices of the nuclear industry in New Mexico and Arizona were not contested because those impacted did not have enough political sway. The domestic nuclear empire grew unchecked (Genay, 2019, pp. 41–42). By comparison, wealthy communities—or those with more political clout—experienced faster action than Navajo sites, for instance, the experience in two cities in Colorado and Utah. In Grand Junction, Colorado, the extensive radioactive pollution led to the establishment of a Superfund program (Rock and Ingram, 2020, p. 20). The city, located in western Colorado, two hundred miles from the Navajo Reservation, was successful in demanding action for cleanup from the pollution left by the Climax uranium mine (Hearing, 2007, p. 166) and is now able to rebrand itself as the "heart of wine country." The city of Moab, Utah, was the "the uranium capital of the world" in the 1950s (Amundson, 2004, pp. 57, 63). After a Superfund cleanup—still underway—to remove a large pile of tailings on its outskirts, the city has become a world class "red rock" tourist site (Amundson, 2004, p. 126).

But the Navajos' experience differed. In 2007, Steve Etcitty, director of the Navajo Nation EPA, brought a sample of radioactive soil from the Navajo Nation to the congressional hearing in Washington, DC. Even though he had chosen a sample with a lower level of radioactivity, he had "to go through extraordinary efforts" to have it pass the Capitol police's security. He made it known that "this is the kind of radioactive dirt that the Navajo people are being exposed to every single day" (Hearing, 2007, p. 122). Addressing the members of the hearing, Dr. Brugge emphatically stated that "none of you would want your children playing in this uranium ore. None of you would permit it" (Hearing, 2007, p. 113).

The correlation between demographics, specifically people of color, or those with lower incomes, and larger impacts from pollutants and other nuisances has been evidenced in the literature (Banzhaf, Ma, and Timmins, 2019). Environmental injustice materializes in "the overwhelming tendency of toxic waste facilities to be located in and near African American, Latino, Asian American, and Native communities" (Voyles, 2015, p. 6). In the case of the Navajo, marginalization—and victimization—are added factors as their lack of fluency in the English language, at least from the 1940s through the 1970s, led to their lack of awareness about the dangers of uranium mining and milling (Brugge and Goble, 2002, p. 1411).

Since 1994, Executive Order 12898 requires federal programs to take Environmental Justice populations (EJ) into account with a view to "address the disproportionately high and adverse human health or environmental effects of their actions on minority and low-income populations" (EO 12898, section 1–101; Grinspoon et al., 2014). Impact studies for projects on federal land must listen to the voices of Indigenous Peoples. At times their requests have been granted, for instance, Cave Rock, Nevada, has been closed to sports climbing, or partially addressed, for example Devils Tower, Wyoming, has instituted a voluntary climbing ban in June when Lakota Sundance ceremonies are being held within view of the tower and climbers must refrain from accessing the climbing routes. As a world class climbing site, Devils Tower attracts visitors from all over the world and, so far, climbers have mostly been compliant. In other cases, a negotiation has taken place, as in Snowbowl, Arizona, where floodlights, that would have allowed nigh-skiing, have been banned. All of these issues, whether it is the lack of consideration for indigenous lives, the protection of their land, their belief systems, and their ceremonies have evolved for the best since the 1970s. Indigenous peoples are no longer invisible, and the United States can be considered as a country that has become more aware of the importance of indigenous rights. However, critics will say that Indigenous Peoples have not been granted much in the face of past dispossession (Berthier-Foglar, 2010).

In the cases discussed in the previous paragraph—Cave Rock, Devils Tower, Mount Taylor—the areas in question were part of the historic land of one or more tribes, but they were not part of a reservation. The difference is important, as it would be very difficult today to coerce a tribe to give up reservation land, although it is held in trust by the United States for the tribes as stated in the treaties. As for federal land, the current legislation grants access rights for tribes and progress has been made concerning tribal input in the management of federal non-reservation land if they can prove historic use.

As of 2023, the geopolitical dynamics emphasize the need for the United States to maintain their energy and raw materials independence. New and growing needs have, once more, directed the focus on the uranium resources of the Four Corners area. But because the pre-1990s mines and mills have not yet been cleaned up, the Navajo are now wary of new projects.

Sacrificial Zone 2: Copper and Clean Energy

New energy needs, especially for clean energy, require massive amounts of copper. This is the premise for the second potential sacrificial zone to be examined, the Resolution Copper Mine, sixty miles east of Phoenix, Arizona.

In 2023, the world is crisscrossed with an extensive network of copper wiring transporting electricity from the producers to the users. Domestic users, like each of us, own an increasing array of devices that function thanks to copper. Life without electronics and battery-powered equipment is difficult to imagine.

Copper is valuable. It can be formed into small wires without breaking, it is a good conductor of electricity and heat, and it is recyclable without any loss of its properties. The energy transition, that is, the switch to clean energy, cannot take place without copper. Solar, wind, and geothermal power are intensive copper users, as are batteries. The worldwide energy transition needs more and more copper: "Copper demand is expected to double by 2035, and continue to grow thereafter" (S&P, 2022, p. 27). Hence the importance of a domestic copper supply for the United States.

Copper is one of the historic industries of Arizona where large swaths of land are pockmarked with mines, open-pit extraction sites, and smelters. In central Arizona, the Resolution Copper mine has been in the planning stages since the discovery, in 1995, of a large body of ore, its apex 0.8 miles below ground, in a sparsely populated area sixty miles east of Phoenix (Briggs, p. 11). It is a low-grade ore, like most of the ore discovered today. With a 1.54 percent grade, it means that for every ton of ore, an average of thirty-one pounds of copper will be obtained (Final Environmental Impact Study, 2021, Volume 1, ES 1.6.1).

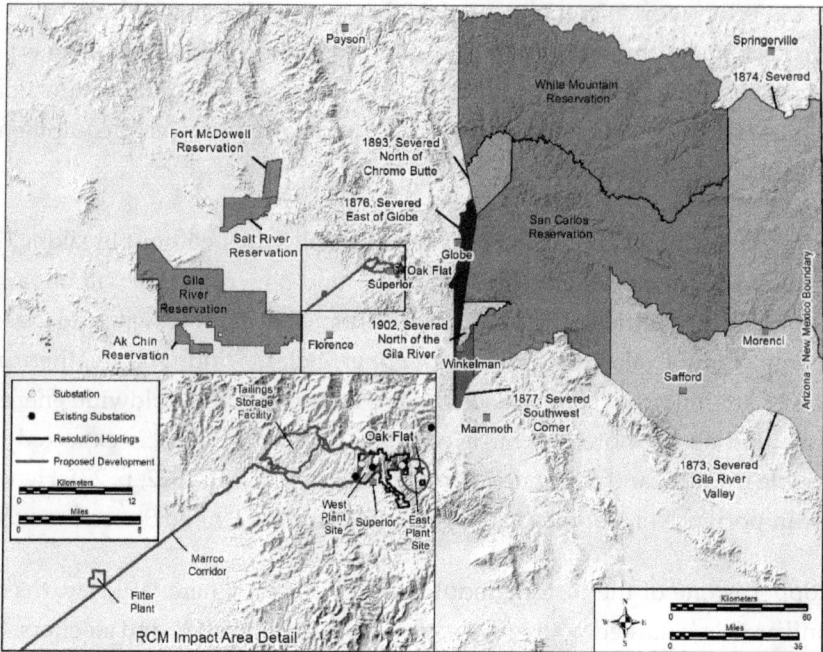

Figure 2.2. *Map of Western Arizona with the reservations surrounding the mining town of Superior: The San Carlos and the White Mountain Reservations (both Apache), and the smaller reservations—Gila River, Ak Chin, Fort Mc Dowell, and Salt River—home to the Pima, Maricopa, and Yavapai. The map also features the reduction in reservation surface between 1873 and 1902, as well as the surface impact of the Resolution Copper Mine (Welch, p. 2).*

The unfolding story is in some ways similar to what happened in Navajo country seventy years ago. It focuses on the acceptability of creating a wasteland to obtain a badly needed resource. However, seventy years ago, environmental impact studies did not exist, and the voices of the local populations were not heard. Moreover, the term *wasteland* was not part of the vocabulary to describe overuse by mines.

The present chapter does not aim to discuss the technical side of the Resolution Copper mine but rather its human impact, and more specifically, its impact on local people. Two main issues come to the fore. One is the specific status of the land above the ore, Oak Flat, a "lush desert riparian area" (Redniss, 2020, p. 18), a National Forest land, withdrawn from mining by President Eisenhower to preserve its pristine setting

(Lovett, 2017, p. 356). The other issue is the staunch opposition from the San Carlos Apache, and more specifically, from their lobby, the aptly named Apache Stronghold, who reject any form of destruction of Oak Flat. They have reclaimed the area's Apache name, Chi'Chil Bildagoteel, as well as its ancestral use and, in 2016, had it listed as a Traditional Cultural Property on the National Register of Historic Places (Final Environmental Impact Study, 2021, Vol. 1, ES 28).

In theory, the Declaration on the Rights of Indigenous Peoples, passed by the UN in 2007 and ratified by the United States, gives the Apache the right to all their ancestral territory and its resources. However, the Declaration is nonbinding, as it would require the entire territory of the United States to be vacated and given back to Indigenous Peoples. While giving back the continent is utopian, mitigation is an accepted practice, and mining companies are now legally required to follow guidelines that were nonexistent during the uranium boom of the Cold War.

Near the mine, twelve communities concerned with the EJ (Environmental Justice) are to be found; eight of them are Native American. All are vulnerable or disadvantaged but the Indigenous communities would also be negatively impacted by their loss of tribal values with a mining operation next to their homeland (Final Environmental Impact Study, 2021, Vol. 1, ES 29).

As for the mine's impact, Resolution Copper plans to use a new type of underground mining called *block caving*, entailing sectional fracturing of the ore from below, underground crushing, and transportation above ground for refining and smelting. The location of an old smelter in the nearby town of Superior, Arizona, will be used. Plans call for a tailing storage facility to be built on a nearby site, yet to be determined. Block caving will not immediately destroy Oak Flat. In the forty years of active mining—*i.e.* during the extraction process—the surface above the mine will gradually collapse, forming—in the words of the impact study—"a large visible crater," approximately one thousand feet deep and two miles wide (Final Environmental Impact Study, 2021, Vol. 1, p. 63).

Before mining in the Oak Flat "withdrawal area" can start, a land exchange must take place giving private land of similar value to the National Forest. It is an accepted practice and Resolution Copper has been buying private parcels in the past decade for that purpose (Lovett, 2017, pp. 355–56). Thus, the mining company offered to exchange 5,460 acres of private land on eight parcels, located in Arizona, for the 2,422 acres of the Oak Flat Federal Parcel (Final Environmental Impact Study, 2021, Vol.1, ES 8). They are ecologically significant and have high aesthetic value. One of the parcels— Apache Leap South End—has an additional high cultural value due to its proximity to the sensitive site of Apache Leap, considered by some to be the location of a mass suicide of Apache warriors refusing capture.

Nevertheless, no land exchange can hide the fact that Oak Flat will be destroyed, and opposition to the mine—to any site of extraction in general—is strong. It is a core problem of the energy-raw-materials nexus, that is, the connected relation between energy and raw materials (Vidal, 2022, p. 8). Energy is needed to sustain the current lifestyle Americans are used to, and raw materials—including copper—are part of the equation.

Wastelanding is adamantly rejected and not only by those who are directly impacted. The 2020 destruction of an aboriginal rock shelter at Juukan Gorge, in Western Australia, by a mining company with the legal authorizations to proceed, led to a worldwide public outcry resulting in the firing of the company's three top officials (Allam and Wahlquist, 2021). The fact that Rio Tinto, the company also owning 55 percent of Resolution Copper, was involved in the Juukan Gorge destruction, confirms activists' belief that global mining companies are essentially "evil" (Nosie, n.d.).

In the United States, the permitting process, the period between the exploration for a mine and the start of exploitation (*i.e.* the actual mining phase) starts with a Draft Environmental Impact Study (DEIS), followed by a period of public input which is then followed by a Final Environmental Impact Study (FEIS).

In the case of Resolution Copper, one criticism leveled at the DEIS by several tribes is the absence of tribal monitors in the archaeological surveys of the areas to be impacted. The 2021 FEIS now includes the findings by representatives of the four Indigenous groups with a link to the sites. Culturally sensitive sites were identified: historic properties, burial sites, sacred places, springs and water resources, and hunting and gathering areas (Final Environmental Impact Study, 2021, Vol. 3, pp. 35–45). During the survey, tribal monitors found sites they had not been aware of, and were greatly distressed by their potential loss. In the Oak Flat parcel, they recorded 594 special interest sites, 523 cultural resources, and 66 natural resources (landforms, water, and mineral resources). Due to the sensitive nature of the findings, only general information is given in the FEIS and exact locations are not provided (Final Environmental Impact Study, 2021, Vol. 3, p. 833).

The spotlight on Oak Flat since the early stages of the permitting process in 2003 (Redniss, 2020, p.18) led to the reappropriation, and resacralization of the area by Apache activists. Ironically, their actions were reinforced and justified by the findings of the archaeological survey sponsored by Resolution Copper, especially since forgotten burial sites were rediscovered.

The ancient land use by the Apache cannot be disputed. There was, however, a long gap in tribal use. The proposed copper mine is in the Globe-Superior area, heavily mined since the 1870s, first for silver, then for copper. It was part of the original Apache homeland, lost during the Apache Wars, a series of armed conflicts between the US Army and various Apache tribes, during the early extractive period from the 1870s to the late 1890s (Sheridan, 2012, pp. 159, 168–71), thus in the 1970s, Oak Flat was no longer considered sacred and in the following years, new mining exploration did take place without protest from the tribe (Miles, 2015). However, with the land exchange under consideration by the lawmakers, Apache activists have manned a camp on several occasions to reclaim ancient use and meaning for the tribe (Lovett, 2017, p. 378).

Major changes have occurred since the late 1970s in the protection of Native American religions and the passing of the American Indian Religious Freedom Act (AIRFA, 1978). With AIRFA and subsequent legislation, the words of the First Amendment, "Congress shall make no law respecting an establishment of religion, or prohibiting the free exercise thereof," came to be viewed through the lens of Native American needs and practices (Berthier-Foglar, 2011, pp. 7–11). For the Indigenous antimining lobby, the destruction of Oak Flat denies their right to practice their religion (FEIS, 2021, Volume 6, p. 14), and no mineral underground is valuable enough to warrant the surrender of Oak Flat.

Meanwhile, the permitting process, started in 2004, involved multiple agencies and departments as well as an input of local populations, indigenous and others, land users and non-governmental organizations. Resolution Copper demonstrated their consideration of local communities with an extensive hiring policy targeting individuals with few qualifications. Scholarships were set up for local students. Community programs have been financed, and a total of $2.3 million has been donated to local communities in 2021 (Resolution Copper, 2022). And to avoid the catastrophic onslaught of negative publicity faced by Rio Tinto after the Juukan Gorge affair, Resolution Copper has proposed remediation efforts to counter every criticism leveled at the proposed mining operation. The sensitive Apache Leap cliff would be saved by stopping the extraction at a distance from the protected area. A stand of Emory Oak—for which Oak Flat is named—would be transplanted, at great cost, to save them from destruction. Burial sites could be relocated. Culturally sensitive sites, located on private property to be transferred to the National Forest, would be protected.

However, any kind of remediation proposal is considered preposterous, even insulting, to the hardliners in the Apache community who refuse to see their historic land transformed into a sacrificial zone. While remediation is the key concept in negotiations with the communities concerned by the Environmental Justice legislation, the difficulty comes from the impossibility of setting a price on noncommercial resources such as clean air and water, silence, and well-being, and ultimately, on traditional religious practices. Less than three months after its publication, in a rare move, the FEIS was rescinded and the consultation process started anew.

No form of negotiation is acceptable to the most vocal lobby against the mine, Apache Stronghold, while at the same time, an antimining lobby is on the rise in the United States[1] (S&P, 2022, p. 86). On the other hand, most communities surrounding the mine, express hope to have jobs close to home (Final Environmental Impact Study, 2021, Vol. 6). Poverty and a high unemployment rate are the main reasons for a choice that has previously been labeled "environmental blackmail" in the context of New Mexican uranium mines (Voyles, 2015, p. 193).

Today, the situation of copper is different. The public outcry for green and clean energy, and for a transition away from fossil fuels, has placed the focus on exploiting the existing copper reserves, and providing wind and solar energy as well as electric cars to a large segment of the US (and the world's) population. While copper is not yet a critical metal, as other sources are found in the United States and the Americas (S&P, 2022, p. 10), it is ironic to remark that some of its uses are related to the demands of a protest movement claiming that inaction, that is, relying on fossil fuels, "takes the future away" from the youth of the world. (United Nations Framework Convention on Climate Change, 2018).

Digging the Earth: An Uncertain Future for Indigenous Peoples

Two distinct narratives characterize the issues surrounding the digging of the earth. They are written by those who want to use the resources found underground and by those who refuse their extraction (Kuletz, 1997, p. xix). A third option, a nonnarrative, is implicitly produced by those who turn a blind eye to the extractive industries, especially if they are located in distant countries with little or no media coverage. The "blind-eye end-users" of raw materials, as they could be called, give de facto support to mines operating without environmental or social considerations, as long as they do so in distant countries. The blatant hypocrisy of the situation is a variant of the "Not In My Backyard" syndrome. The Biden administration seems torn between striving to protect nature, including Indigenous land, and avoiding the dependence on foreign resources (White House, 2022). In discussions about mining, the mining law, and the permitting process to open new mines, the outcome remains undecided.

In the field, the movement to protect Oak Flat attempts every possible way to stall the opening of the Resolution Copper mine (Krol, 2022). In matters of low carbon energy, when domestic uranium supplies are discussed, Navajo voices are heard opposing the reopening of mines in the context of the failed cleanup from the previous mining phase (Dooley, 2008).

The future remains uncertain. While Indigenous Peoples in the United States may have a more acceptable protection against the excesses of the extractive industries; in other countries, they often face administrations that offer little sympathy or outright indifference. Public support in favor of Indigenous Peoples often disregards the fact that the last word is with the end users of the raw materials extracted from the earth. Their uses range from the critical to the mundane, from communications infrastructure to mere gadgets. Drawing the line between critical and futile uses is up to the citizens, the end users, the consumers. Choices will have to be made. None is harmless.

To achieve a decarbonized energy supply, copper from Apache country will be needed, possibly also uranium from Navajo country. Clamoring for action on the part of governments, while refusing policies imposed from above, will not solve the present crisis. Voluntarily redirecting one's personal consumption of raw materials is an effective action, but frugality in materials use is rarely mentioned as its implementation by manufacturers may be perceived by the consumer as undue cost-cutting.

Issues surrounding uranium and copper are not isolated case studies and similar developments can be found in the context of lithium and rare earth elements. The projected Thacker Pass lithium mine in Nevada could become the largest producer of this critical material in the United States as lithium, today mainly produced in China, is the main component of the batteries for electric vehicles. The Paiute refuse to see their land blighted by the mine (Rothberg, 2021). The Mountain Pass mine for rare earth elements in California, previously shut down in the face of a pollution scandal, was reopened to become the sole provider of most rare earth elements for the United States (Xie, 2020).

The hunger for raw materials seems unstoppable and it foreshadows an alarming situation for Indigenous Peoples of the world, especially in countries where governments don't listen to their voices.

Notes

[1] The anti-mining lobby in the United States calls for a "revision of the regulations to create an environment that is even less conducive to mineral extraction than exists already" or even a prohibition of all forms of hard rock mining on federal lands, although this is less likely to succeed (S&P, *The Future of Copper*, p85-86). Both policies may simply move mining operations out of sight to countries with little public scrutiny, and into the hands of companies with no accountability.

References

ACHRE (1995). *Advisory Committee on Human Radiation Experiments. Final Report.* Washington, DC: U.S. Government Printing Office (October 1995). Available at: https://archive.org/details/advisorycommitte00unit

American Indian Religious Freedom Act of 1978 (AIRFA). Available at: 42 U.S.C. §, 1996.

Allam, L., Wahlquist, C. (2021). A year from the Juukan Gorge destruction, Aboriginal sacred sites remain unprotected, *The Guardian*, May 23. Available at: https://www.theguardian.com/australia-news/2021/may/24/a-year-on-from-the-juukan-gorge-destruction-aboriginal-sacred-sites-remain-unprotected

Amundson, M. (2004). *Yellowcake Towns: Uranium Mining Communities in the American West.* Boulder, CO: University Press of Colorado.

Arnold, C. (2014). Once Upon a Mine. The Legacy of Uranium on the Navajo Nation, *Environmental Health Perspectives*, 122(2), pp. A45-A49.

Banzhaf, S., Ma L., Timmins, C. (2019). Environmental Justice: The Economics of Race, Place, and Pollution. *The Journal of Economic Perspectives*, 33(1).

Berthier-Foglar, S. (2010). Mythic Mountains in the United States: Sacredness and the Law. In Besson, F. (Ed.) *Mountains Figured and Disfigured in the English-Speaking World*, pp. 634-645. Newcastle upon Tyne, UK: Cambridge Scholars Publishing.

Berthier-Foglar, S. (2011). Indigenous Claims and Uranium Mining on Mount Taylor, New Mexico, USA. *The Journal of the International Association of Inter-American Studies. Indigenous America – América Indígena*. 4(2) Available at: http://www.interamerica.de/

Briggs, D. F. (2015). Superior, Arizona: An Old Mining Camp with Many Lives. *Arizona Geological Survey*, Report CR-15-D, pp.1-13. Available at: repository.azgs.az.gov

Brugge, D. (2000). *Memories Come to Us in the Rain and the Wind. Oral Histories and Photographs of Navajo Uranium Miners and Their Families.* Jamaica Plains, MA: Red Sun Press.

Brugge, D., and Goble, R. (2002). The History of Uranium Mining and the Navajo People. *American Journal of Public Health*, 92 (9), pp. 1410-1419.

Brugge, D., Benally T., Yazzie-Lewis, E. (Eds.) (2006). *The Navajo People and Uranium Mining.* Albuquerque, NM: University of New Mexico Press.

Community Profile for Superior (2022). Arizona Commerce Authority. Available at: https://www.azcommerce.com/a/profiles/ViewProfile/123/Superior/

Dawson, S., Madsen, G. (2011). Psychosocial and Health Impacts of Uranium Mining and Milling on Navajo Lands. *Health Physics,* 101(5), pp. 618–625.

Dooley, E. (2008). Navajo Fight Uranium Comeback. *The Beat | Environmental Health Perspectives,* 116 (5). Available at: https://ehp.niehs.nih.gov/doi/10.1289/ehp.116-a200b

Eichstaedt, P. (1994). *If You Poison Us. Uranium and Native Americans.* Santa Fe, NM: Red Crane Books.

Environmental Protection Agency (2016). *Abandoned Uranium Mines on or near Navajo Nation,* Map. https://www.epa.gov/navajo-nation-uranium-cleanup/abandoned-mines-cleanup

Environmental Protection Agency and Navajo Nation Superfund (2017). *Contaminated Structures Program,* EPA Region 9. August 2017. https://www.epa.gov/navajo-nation-uranium-cleanup/addressing-uranium-contaminated-structures

Executive Order: Federal Actions to Address Environmental Justice in Minority Populations and Low-Income Populations, (1994). *Federal Register,* 59(32), February 16. Available at: https://www.archives.gov/files/federal- register/executive-orders/pdf/12898.pdf

FEIS. *Final Environmental Impact Study. Resolution Copper and Land Exchange.* USDA. Forest Service. Tonto National Forest. January 2021. https://www.resolutionmineeis.us/documents/final-eis (accessed 20 January 2021).

Genay, L. (2019). *Land of Nuclear Enchantment. A New Mexican History of The Nuclear Weapons Industry.* Albuquerque, NM: University of New Mexico Press.

Grinspoon, E., Schaefers, J., Periman, R., Smalls J., Manning C., Lo Porto, T. (2014). *Striving for Inclusion. Addressing Environmental Justice for Forest Service NEPA.* Washington D.C.: U.S. Forest Service. June. Available at: https://www.resolutionmineeis.us/sites/default/files/references/grinspoon-et-al-2014.pdf

House Hearing, 110 Congress (2007). The Health and Environmental Impacts of Uranium Contamination in The Navajo Nation. *House of Representatives, 110th Congress, 1st session,* October 23. Available at: https://www.govinfo .gov/content/pkg/CHRG-110hhrg45611/html/CHRG-110hhrg45611.htm

Johannsen, B. (2013). *Encyclopedia of the American Indian Movement (Movements of the American Mosaic).* Santa Barbara, CA: Greenwood.

Krol, D. (2022). Federal appeals court denies Apache Stronghold's bid to stop a copper mine at Oak Flat. *Arizona Republic* June 28. https://www.azcentral.com/story/news/local/arizona/2022/06/28/9th-circuit-court-denies-   appeal-stop-copper-mine-oak-flat/7750393001/

Kuletz, V. (1997). *The Tainted Desert. Environmental and Social Ruin in the American West.* Abingdon, UK: Routledge.

Lovett, K. (2017). Not All Land Exchanges Are Created Equal: A Case Study of the Oak Flat Land Exchange. *Colorado National Resources, Energy & Environmental Law Review,* 28(2), pp. 355-386.

Matthews, Washington (1897). *Navaho Legends.* Salt Lake City: University of Utah Press. 1994 reprint.

Miles, D. (2015). Oak Flat is a sacred site? It never was before. *AZ Central,* July 23. Available at: https://www.azcentral.com/story/opinion/op-ed/2015/07/23/oak-flat-sacred/30587803/

Navajo Department of Health (2022). *Uranium Workers Program.* The Navajo Nation. Available at: https://ndoh.navajo-nsn.gov/Department/Division-of-Public-Health-Service/Uranium-Workers-Program

Nosie, W., Sr. (2023). The holy places are rumbling at what is happening in the world. *Poor People's Campaign.* Available at: https://www.poorpeoples campaign.org/we-cry-power/wendsler-nosie/

Pasternak, J. (2006). A Peril that Dwelt Among the Navajos. *Los Angeles Times,* Nov 19, Available at: https://www.latimes.com/archives/la-xpm-2006-nov-19-na-navajo19-story.html

Pasternak J. (2011). *Yellow Dirt: A Poisoned Land and the Betrayal of the Navajos.* New York: Free Press.

Radiation Exposure Compensation Act (1990). *Congressional Research Service,* Public Law 101-426-Oct. 15,1990

Redniss, L. (2020). *Oak Flat: Fight for Sacred Land in the American West*. New York: Random House.

Resolution Copper (2022). Corporate Communication. https://resolutioncopper.com/supporting-our-community/

Rock, T., Ingram, J.C. (2020). Traditional Ecological Knowledge Policy Considerations for Abandoned Uranium Mines on Navajo Nation. *Human Biology*, 92(1), Winter 2020, pp. 19–26. Available at: DOI: 10.13110/humanbiology.92.1.01.

Rothberg, D. (2021). Amid Plans to Mine Lithium in Rural Nevada, Indigenous, Rural Communities at Center of the Energy Transition. *The Nevada Independent*, June 22. https://www.nnbw.com/news/2021/jun/22/amid-plans-mine-lithium-rural-nevada-indigenous-ru/

Sheridan, T. E. (2012). *Arizona, A History*. Tucson, AZ: University of Arizona Press.

S&P Global (2022). *The Future of Copper. Will the looming supply gap short-circuit the energy transition?* Available at: https://www.spglobal.com/market intelligence/en/mi/info/0722/futureofcopper.html

United Nations Framework Convention on Climate Change (2018). *COP 24*. Greta Thunberg. Available at: https://unfccc.int/sites/default/files/resource/COP24_HLS_ENGO.pdf

USDA Forest Service, Tonto National Forest. (2021). *Resolution Copper and Land Exchange Environmental Impact Study*, January 2015. Available at: https://www.resolutionmineeis.us/documents/final-eis

Vidal, O., Goffé, B., Arndt, N. (2013). Metals for a Low Carbon Society. *Nature Geoscience*, 6, pp. 894-896.

Vidal, O., Le Boulzec, H., Andrieu B., Verzier, F. (2022). Modelling the Demand and Access of Mineral Resources in a Changing World. *Sustainability*, 14(1). Available at: https://doi.org/10.3390/su14010011.

Voyles, T. B. (2015). *Wastelanding: Legacies of Uranium Mining in Navajo Country*. Minneapolis,MN: University of Minnesota Press.

Welch, J. R. (2017). Earth, Wind, and Fire: Pinal Apaches, Miners, and Genocide in Central Arizona, 1859-1874. *SAGE Open*, December 2017, pp.1-19. Available at: https://journals.sagepub.com/doi/10.1177/2158244017747016

White House (2022). *Fact Sheet: Securing a Made in America Supply Chain for Critical Minerals*, February 22, 2022. Available at: https://www.whitehouse.gov/briefing-room/statements-releases/2022/02/22/fact-sheet-securing-a-made-in-america-supply-chain-for-critical-minerals/

Xie, J. (2020). California Mine Becomes Key Part of Push to Revive US Rare Earths Processing. *VOA*, Dec. 31. Available at: https://www.voanews.com/a/usa_california-mine-becomes-key-part-push-revive-us-rare-earths-processing/6200183.html

Zoellner, T. (2010). *Uranium: War, Energy, and the Rock that Shaped the World.* New York: Penguin Books.

Chapter 3
Mining in Mexico and the Land Defenders Movement in Chiapas

Lynn Holland

. . . each body a lion of courage, and something precious to the earth.
——Mary Oliver

Even in his own day, English mining magnate Cecil J. Rhodes was known for his unabashed embrace of imperialism. His remark about Mexico's potential as the world's source of mineral wealth is all the explanation we need of imperialist ambition in the late nineteenth century. At just twenty-four years of age, he proclaimed, "I am not blind to the unison of opinion, expressed by scientists and experts, that Mexico will someday furnish the world; that from her hidden vaults, her subterranean treasure houses, will come the gold, silver, copper and precious stones that will build the empires of to-morrow and make future cities of this world veritable New Jerusalems" (Osterheld, 1916, p. 3).

To build these empires of tomorrow and future cities of the world, imperialist countries would need unimpeded access to the natural resources of colonized countries. Existing communities of Indigenous People in these regions therefore would have to be removed, pacified, or otherwise "dealt with." This would only be accomplished through the forceful imposition of a system of individual property rights at the expense of the pre-existing notion of communal land and decision making. Referring to Africa, young Cecil Rhodes rationalized these genocidal tactics by asserting the idea of Anglo-Saxon racial superiority over Native people, "the most despicable specimen of human being" (Carr, 1964, p. 76).

What follows is a short history of mining in Mexico and its impacts in the resource-rich southern state of Chiapas. Much as it has in the past, the extraction of natural resources today allows for the amassing of some of the greatest fortunes in the world. At the same time, it is responsible for poisoning and dispossessing Indigenous and other rural people who depend on the land for their survival. In response, many of these communities are choosing to defend their land against the advance of mineral mining while offering the world an ecological alternative to Rhodes's vision of power and wealth.

Mining and Development in Mexico

In the conventional view of development, the path to modernization for poor countries lies in their exposure to Western values. Known as "modernization theory," the proponents hold that "traditional" values associated with subsistence agriculture pose an impediment to development, which is best overcome through foreign aid and investment in the less developed country. Through this transfer of capital from more developed countries, the less developed country is able to take on the political, social, and economic characteristics of modern Western societies. This includes increasing levels of mass consumption (Rostow, 1961; Sachs, 2006).

Critics of modernization theory note that, while a consumer class does in fact emerge in these cases, it is an *elite* class with a relatively high standard of living. Meanwhile, much of the rest of society becomes more impoverished. Known collectively as dependency theorists, these critics divide the world into capital-rich or "core" countries and resource-rich or "peripheral" countries. As investment capital flows from core to periphery, the governments of peripheral countries become less responsive to the needs of their own constituents and more responsive to investors in the core. On the other hand, as investment slows, governments in the periphery become less outwardly dependent and more attuned to the demands of their own constituents. It is during these periods of greater state autonomy that internal reform becomes possible (Prebisch, 1962; Frank, 1967).

The mining industry in Mexico provides a clear example of these cycles of boom and bust. As the demand for mineral resources in core countries rises, partnerships are formed between foreign and domestic capital, which are capable of placing tremendous pressure on the peripheral governments. These dynamics have played out from one era to the next in Mexican history with profound implications for Indigenous Peoples.

During the colonial period, the Spanish founded major cities such as Taxco, Zacatecas, Guanajuato, and Potosí where large deposits of silver were discovered. These mines came to account for 90 percent of New Spain's total exports during the colonial period transforming Spain into a wealthy and powerful empire. Mexican silver, dispensed throughout the world, supported the international trading system and the process of monetization in Europe and the Americas (Mexico Business News, 2013). At the same time, colonization left a legacy of hostility between a conservative landowning elite of Spanish descent and increasingly landless Indigenous populations.

The years from 1876 to 1911, known as the Porfiriato, brought another surge in foreign investment as then President Porfirio Díaz made mineral exports a centerpiece of economic growth. As laws were passed to allow foreign companies easy access to landownership, Indigenous communities, often lacking title, were deprived of their communal lands. "Barbed wire would go up soon after, fencing off what the villagers had always regarded as theirs, and often depended on for their survival" (Easterling, 2012, p. 17).

As the Mexican Revolution approached, landlessness became the central issue for the three-fifths of the population who were Indigenous (Easterling, 2012, p. 23). Accordingly, Indigenous groups fought to legitimize the ejido, a form of collective land ownership derived from "an Indigenous system of land tenure that had survived the four centuries following the arrival of the Spaniards in 1519" (Kelly, 1994, p. 545). Emiliano Zapata, a farmer from Morelos, took a leading role in this movement. Under the slogan "land and liberty," Zapata issued Plan Ayala, a call for the redistribution of land to those who work it and national oversight of agrarian reform hearings (Grubacic and O'Hearn, 2016, p. 113).

Both in life and in death, Zapata's influence has been legendary. He is credited with promoting the 1915 decree that reversed the nineteenth-century division and alienation of communally owned lands and established hearings on land reform (Stone, 1978). Two years later, Articles 27 and 42 of the new Mexican Constitution declared all land, water, and mineral resources to be the property of the people of Mexico and gave the government the authority to expropriate land from the largest estates for redistribution to eligible rural communities (Kelly, 1994).

When the fighting came to an end in 1920, "Mexico had a populist agrarian law based on Zapata's ideas and designed to meet the needs of the nation's indigenous population" (Carrigan, 1995, p. 79). The first ejidos granted to Indigenous Peoples were in the Lacandon rainforest in Chiapas in the 1930s, an area which had been abandoned by pre-Columbian Maya groups centuries before. The migrants who arrived from other regions in Chiapas represented a number of Indigenous groups including Tzeltal-, Ch'ol-, and Tojolobal-speaking Mayans. In order to form cohesive ejido communities, these migrants struggled to adjust their customs and learn one another's languages while also learning how to survive on unfamiliar terrain. In time, these struggles paid off in the development of their own semiautonomous governing structures which would become the base of the Zapatista movement in the decades that followed (Grubacic and O'Hearn, 2016, p. 120).

The Rise of Neoliberalism and the Zapatista Revolution

By the late 1980s, a new era of economic expansion was underway. This was prompted by the ideology of "neoliberalism," a repackaging of modernization theory that promotes a shift in power from national governments to transnational capital. Neoliberal economists argued vigorously that cutting back on government services and regulations would improve the economy by generating new levels of foreign investment. Investors in the US and Mexico along with international banks and the US government pushed hard for these cutbacks in an effort to roll back the hard won achievements of the Mexican revolution. In line with these efforts, the International Monetary Fund (IMF) offered new lines of credit conditioned on the cutbacks and the privatization of natural resources (Stoltenborg and Boelens, 2016).

Despite stiff opposition, the Mexican Congress obliged. In 1992, Article 27 was revised to allow the privatization and sale of ejido land, effectively ending the government's obligation to allocate land to the rural poor (O'Toole, 2017). A law was also passed to allow the buying and selling of ejido lands. Publicly owned mining companies were sold off often at prices well under market value, and foreign-owned mining companies were allowed full ownership of natural resources. Taxation levels were reduced, and the mining concession period was extended from twenty-five to fifty years (Tetrault, 2014). In 1994, the implementation of the North American Free Trade Agreement (NAFTA) between the US, Mexico, and Canada allowed private companies to bypass national law and take their disputes to the World Bank's International Centre for the Settlement of Investment Disputes (ICSID) instead. Yet this same right was denied to those communities whose lands had been contaminated by the mining process (Stoltenborg and Boelens, 2016). By the end of the decade, public hearings were no longer required for the granting of mining concessions, leaving local residents uninformed about the environmental and other adverse effects of mining projects in their communities.

As these legal and political changes took effect, mining investments soared, reaching a peak of over $8 billion in 2012 (Reuters, 2022). In the years since, Mexico continues to lead the world production of silver while placing ninth in the production of gold, and tenth in copper. It is also among the top ten producers of barite, zinc, lead, gypsum, molybdenum, and other minerals (Morales, A., 2017). The capacity for mining has greatly increased with the development of powerful new mining technologies. Today companies can destroy entire mountaintops, strip forests, and extract flecks of ore so small they cannot be seen by the naked eye. As a result, mining companies have extracted approximately twice as much gold and one and a half times as much silver from Mexico in the last fifteen years as was extracted in the entire three-hundred-year period of colonization (Tetrault, 2015).

Mining has also contributed to many of the great fortunes of Mexico including those of Carlos Slim, owner of Minero Frisco, and Germán Larrea

Mota-Velasco, CEO of Grupo México, Mexico's largest mining company. Moreover, it has helped to put mining companies like US-owned Newmont and Canadian-owned Teck Resources among the largest and wealthiest in the world (Gobierno de México, 2019). At the same time, Chiapas has grown more polluted and the people more impoverished.

Much as it had during the Porfiriato, this era of economic expansion was accompanied by increasing levels of poverty among Indigenous and other rural communities. With the goal of challenging the power of transnational capital and bringing this poverty to an end, the Zapatista Army of National Liberation (EZLN) was founded by three Indigenous and three non-Indigenous individuals in 1983. Among the non-Indigenous members who joined early on were students who had been involved in movements for the urban poor in Monterrey. Still others had participated in rural rebellions in Guerrero during the 1970s (Carrigan, 1995, p. 82). In 1972, members of the National Liberation Front (FLN), a left-wing guerrilla organization, had purchased a ranch near San Cristóbal in Chiapas. After it was nearly wiped out by the Mexican army in 1974, surviving members joined the founders of the EZLN (Wager and Schultz, 1995, p. 11).

The Catholic Church also played a role in the shaping the EZLN. In 1974, the city of San Cristóbal commemorated the 500th anniversary of the birth of Bartolome de las Casas, a Spanish colonial Dominican priest who had appealed to the king of Spain to enact laws against the exploitation of Indigenous people. The event produced a wave of new grassroots organizations working on behalf of Indigenous and other rural communities in Chiapas and the surrounding area.

Among their supporters were adherents of Latin America's liberation theology as set out at the Latin American bishops conference in 1968 (Wagner and Schultz, 1995, p. 11).

Chief among them was Don Samuel Ruíz Garcia, bishop of the Diocese of San Cristóbal home to Mayan communities speaking many different languages. On visiting these communities by mule, Bishop Ruiz developed a deep appreciation for their unique customs and sympathy for their impoverishment (Carrigan, 1995, p. 81). In their early encounters, the left-wing revolutionaries including FLN members found themselves at odds with Bishop Ruíz and the proponents of liberation theology. While they had planned on an armed uprising, the followers of Bishop Ruíz insisted on a peaceful transition through democratic mobilization.

A surge in violent ejido land seizures by cattle ranchers in the late 1980s, however, increased support among Indigenous youth for the creation of armed self-defense units (Carrigan, 1995, p. 82). In 1986, the EZLN entered the first Indigenous community at the invitation of local leaders and by 1989, the organization had over 1300 armed members (Nuijten and Gemma, 2000, p. 85). The 1992 revision of Article 27 allowing the privatization and sale of ejido land and measures to allow foreign owned companies full ownership of natural resources provided further incentives for an armed rebellion.

By 1993, the Lacandón Forest had become a solid base of support for the EZLN and the Indigenous ejidal unions, their essential governing structures. In their first declaration in that year, EZLN leaders presented themselves as the heirs of Villa and Zapata, carrying on the historic struggle for national sovereignty and the rights of the poor (Grubacic and O'Hearn, 2016, pp. 108–76). In another communiqué, NAFTA, faulted for the constitutional revisions which had ended land reform, was deemed a "death certificate for the Indian peoples of Mexico" (Harvey, 1994). Accordingly, EZLN planned an uprising for January 1, 1994, the very day NAFTA was to go into effect.

During the uprising, which lasted twelve days, Indigenous rebels took over public buildings in six major municipalities in Chiapas including San Cristóbal and Ocosingo, along with hundreds of ranches. In a novel move, EZLN leaders turned to the internet to broadcast the conflict aimed at building an international network of supporters[1]. Once peace had been negotiated by Bishop Ruíz, the EZLN withdrew from the occupied civic centers and shifted to a strategy of nonviolent self-defense. In this, they had the backing of numerous local nongovernmental organizations (NGOs).

Two years later, the EZLN and the federal government signed the San Andrés Accords, a peace agreement calling for autonomy for Indigenous communities in Chiapas and a restoration of Article 27 to its original form. The Accords covered five general points of agreement with respect to the Indigenous population in Chiapas: respect for the diversity of that population, conservation of the natural resources owned by that population, the right of consultation by Indigenous peoples regarding public expenditures, the right to autonomy in decision-making regarding development and administration and judicial matters, and the autonomy of Indigenous People within the framework of the state (Organización Internacional del Trabajo, 1996). Civic leaders, academics, and experts were invited to attend the signing as guests and advisers to EZLN (Hernandez, 2010).

The federal government largely ignored the Accords, however, and continued to deploy military as well as paramilitary forces to the autonomous region in Chiapas where they waged a campaign of harassment and sporadic killings. In a particularly horrific attack in 1997, paramilitaries massacred forty-five Tzotzil men, women, and children, including four pregnant women in the village of Acteal, whom they suspected of supporting the EZLN. In fact, the victims belonged to Las Abejas de Acteal, a pacifist Christian organization opposed to violence. Seventy-nine individuals were subsequently convicted but were later released by order of the Supreme Court (Varzi, 2020).

In 2003, the EZLN announced that it was adopting a model of autonomy based on the establishment of five "Good Government Councils" or caracoles. Seven more such councils were established in 2019. Within the caracoles, schools, clinics, and local economies such as coffee growing, bee keeping, handcrafts, and a growing tourist sector emerged. Since then, the caracoles have met with significant challenges. Their services often compete with those of pre-existing government institutions. (Tilly and Kelly, 2006). Funding for services is limited, leading to an overreliance on sympathetic NGOs for financial support. In addition, members have complained of persistent patriarchal tendencies and other internal inequalities (Rebrii, 2020).

At the same time, the education, health care, and the judicial system have all garnered considerable approval from community members. Teachers are typically young adults or "education promoters" who develop the curriculum in consultation with the communities they serve. As such, it invariably includes local Indigenous history, culture, and language among other subjects. For older students, the Indigenous Center for Comprehensive Training in San Cristóbal (CIDECI), born out of the land rights movement in 1989 and inspired by Zapatista thought, serves students from ages 12 to 25.[2] CIDECI commits to building strong autonomous indigenous communities through education in history, art, humanities, civics, agroecology, and numerous trades suited to local economies (In Motion, 2005).

The Zapatista justice system is especially notable for the respect it receives from both EZLN supporters and those outside the community. "It is free of charge, conducted in indigenous languages, and is known to be less corrupt or partial compared to governmental institutions of justice." (Rebrii, 2020). The system favors a restorative process, which aims to repair harm and facilitate remorse in the offender, over a punitive approach. Those found guilty are permitted to return home to cultivate their fields so that the family continues to be cared for.

While in Chiapas in 2016, I asked an indigenous tour guide how he felt about the EZLN. Had there been improvements? He replied that indigenous people had not actually come that far. They were still impoverished, there was infighting among the different communities, and the Mexican army and paramilitary groups were a source of constant harassment. "But things have changed," he said. "Before the revolution, we had to cross the street if a white person was coming. Small children thought it was ok to grab the hat from an Indigenous farmer and taunt him with it. We have pride in who we are now. We believe in ourselves."

Land Defenders in Chiapas. The People's Front for the Defense of Soconusco

From the colonial era to the Porfiriato to the current age of neoliberalism, Indigenous peoples have had to fight for their land against a succession of profiteers. And while mining has played a part in each of these incursions, recent advances in technology have made it ever more destructive to the environment, the health of local residents, and their ability to make a living from the land. The largest concentrations of Indigenous language speakers reside in the southernmost states, which are both rich in metals and other resources and extremely poor in terms of the standard of living. Among them is Chiapas, where 28 percent of the population speak an Indigenous language, the second largest percentage of any state after Oaxaca (Cuentame de México, 2020). And while rich in natural resources, Chiapas is the poorest state in Mexico with 76.4 percent living in poverty (WorldAtlas, n.d.).

Nearly 20 percent of the territory in Chiapas is now under mining concession. By 2019, 140 open-pit mines had been registered with the Ministry with permits allowing control for up to forty years and extensive access to the local water supply. One of the most commonly mined minerals in Chiapas is titanium, used in cosmetics, toothpaste, paint, metal instruments, firearms, computers, and other electronic devices. Gold, silver, zinc, and iron are also mined. Among modern mining techniques, open-pit mining is especially destructive. This procedure is used where minerals are located close to the earth's surface and starts with the carving out of large craters and flattening of mountaintops. Dynamite is used to open up the earth causing the release of arsenic into the air as dust, while the practice of deforestation creates massive ecological disruption throughout the area (Salvemos Cabana, 2011).

Open-pit mines also require enormous amounts of fresh water, which is pumped for drilling, grinding ores, and other tasks, depleting local lakes and streams which residents depend on for fishing, farming, and household use. Small mines typically consume 66,000 gallons of water per hour while large ones consume between 260,000 and 780,000 gallons per hour, a rate that often leaves local farmlands without enough water to meet their needs (Ciacci, 2021). Toxic chemicals such as cyanide and arsenic are used to draw microscopic flecks of ore from the earth leaving tons of hazardous waste rock behind. The waste rock is then deposited in tailings ponds lined with leach pads which can crack during a heavy rainfall or seismic activity. When this occurs, poisons from the tailings ponds enter the rivers, contaminating ground water and soil and creating potential harm for local residents, their livestock, and surrounding wildlife (Holland, 2018).

In Mexico, government documents related to mining concessions are known for omitting the various harms to the environment and human health that are caused by mining. Many Indigenous and other rural peoples who live near mines in Chiapas are coping with respiratory problems, skin rashes, gastrointestinal disorders, and rising rates of cancer, even in teenagers and young adults (Mendoza, 2015). Livestock as well as wild animals have been sickened from drinking contaminated water, schools of fish are dying off, and farmers are watering their crops with contaminated water. These conditions make it increasingly difficult for local residents to make a living from traditional occupations in fishing and farming. Some are forced to give up, sell their land (often to the mining companies), and migrate elsewhere.

In taking action against mining, Indigenous communities have benefited from their experience in the Zapatista movement, which includes participation in self-governance, working with NGOs, building solidarity networks, and the understanding that they have rights. All have provided them with tools for mounting an effective resistance.

In choosing to defend community land against mining, tactics can include direct action such as roadblocks, establishing community-wide referenda, seeking assembly resolutions and public statements against mining, engaging local, state, and national officials as well as the media in discussion over the issues, and taking legal action (Ciacci, 2021). Communities in resistance also reach out or respond to university students, professors, legal clinics, NGOs, journalists, and other supporters in building networks of support. While there are exceptions, the majority of their actions are peaceful.

Acacoyagua and neighboring Escuintla in the region of Soconusco are situated near three titanium mines. A study by the Chiapas state government showed that Casas Viejas mine was in fact the cause of pollution in the waters of the Cacaluta, the main river flowing through the region. This river, which supplies water to households in the area, is untreated by any sewage system thus exposing people to toxins in their own homes. Yet the Mexican government has largely ignored these findings (Ciacci, 2021).

In Acacoyagua, a community of seventeen thousand people learned about the harms of mineral mining the hard way. In 2006, mining companies began exploring in the area and six years later, the first permits were issued. Meanwhile, local residents were given no information nor opportunity for consultation regarding their preferences as a community. The region soon had over a dozen active mines extracting gold, silver, lead, zinc, iron, and titanium (Ciacci, 2021). Within a few years, the community detected a rise in serious health problems among their members including rashes, eye irritation, and gastrointestinal ailments. With the help of two local medical professionals, the community found that the cancer death rate in the area had risen from 7 percent in 2005 to 22 percent in 2015 (Ciacci, 2020). These cases included stomach, testicular, and, especially, liver cancer and were occurring with alarming frequency even in children and young adults. While there, I learned from a local doctor that these illnesses were believed to be the result of drinking, washing clothes, and bathing in local rivers affected by toxic and radioactive particles used in the mining process.

Wildlife was also affected. Fishermen in the community reported declines in the fish population in some areas and deformed fish that could not be sold at market in others. River prawns and other types of marine life were also disappearing. Cattle were sickened after drinking from the river. While some members of the village had sold their land and left the community, I heard that younger people were turning to drug dealing as a way to make a living, and crime was on the rise.[3]

In 2015, forty communities including Acacoyagua and Escuintla came together in the People's Front for the Defense of Soconusco (FPDS), a citizens' organization for the defense of land against mining (LadoB, 2019). As FPDS states, the extractive model of development "damages life, nature, the collective rights of the peoples, the social fabric, the bio-cultural patrimony and our spirituality" (Chiapas Support Committee, 2018). With this, FPDS joined over two thousand such organizations throughout the country, thirty municipalities in Chiapas as well as Guerrero, and the states of Oaxaca and Puebla along with more than eighty ejidal and communal property acts which have declared their territories free of mining (Bessi, 2015).

Figure 3.1. *Members of Frente Popular en Defensa de Soconusco (FPDS) meeting with visitors at the site where they blocked mining equipment from proceeding to the mine in 2018.*

Since then, FPDS has led an effort to close down twenty-one mining concessions in the region. The community has also focused on the impacts of runoff from the mines in La Encrucijada, a biosphere reserve on the coast known for its complex ecosystems, mature mangrove forests, and hundreds of species of mammals, birds, reptiles, and amphibians. In appealing to state and local governments for support, members have found officials to be secretive about who actually owns the mines, and that agents often pass the buck to other agencies making accountability all but impossible (Soberanes, 2017). What they did learn, however, is that many concessions "have changed owners more than once. This is very common in the mining industry, where activities usually start with prospecting and exploration projects in the hands of small or medium-sized national companies, which are later sold to larger investors, either national or transnational, once it is determined that there are enough metal resources to warrant mining" (Ciacci, 2021). Larger mining companies in particular tend to establish subsidiaries through complex legal arrangements that make it difficult to trace ownership.

Mining companies will also court the community they plan to venture into with offers of food, jobs, money, cars, school supplies for children, and offers to buy land. They can be overtly deceptive as well, denying the health hazards and adverse environmental effects related to mining, even claiming that their projects are "environmentally sustainable" (Dilge, 2022). As FPDS responds, "We are not aware of any sustainable mining project, none that does not harm the land, water and health, nor is there any experience where companies have mitigated the impacts of mining upon closure. It seems naive and tricky to us to believe that mining activity will provide us with development and economic gains without polluting or impacting the environment and our health" (REMA, 2019).

With help from nonprofits, FPDS made another important discovery. For a mining project to be approved, a company had to obtain at least twenty-eight permits from different levels of government. The power to issue one of these permits belonged to the municipal council (Morales, Y., 2022). By pressuring the council to withdraw the permits, FPDS was able to shut down the Cristina, El Bambú, and Casas Viejas titanium and ilmenite mines, and secure an official order from the Federal Attorney for Environmental Protection (PROFEPA) (Chiapas Paralelo, 2020). At the same time, the communities carried out direct action against the mines. With just a rope and a rotating guard of eight or ten people, they were able to prevent trucks and machinery from reaching the mine. On my third visit to Acacoyagua in 2018, community members proudly led my delegation to the river which, once polluted, had since become clear and was teeming with small fish.

In response to the suspensions, mining companies Obras y Proyectos Mazapa SA de CV and Grupo Male SA de CV, both with local claims, formed their own organization called the "Regional Mining Committee". Through this committee, mining interests actively seek to convince members of the community to support the reopening of the Cristina Mine through door-to-door campaigns and offers of money in return for support. FPDS has responded by maintaining its campaign to keep Acacoyagua a mining free municipality (Chiapas Paralelo, 2020; Morales, 2022).

The approach taken by the FPDS combines community-centered action and the support of local non-profits with direct confrontation with the mining companies and local government officials that support them. In other cases, Indigenous groups have combined community action with a more institutional approach. In 2020, for instance, the Nahua Tecoltemi community in the state of Puebla took their case against a Canadian gold and silver mining company to the Mexican Supreme Court with support from the Secretariat of Environment and Natural Resources (SEMARNAT). In a victory for the Nahua Tecoltemi, the Supreme Court held that, in granting the concessions, the government had violated the Indigenous and Tribal Peoples Convention Declaring the Right to Free, Prior, and Informed Consultation and Consent, an international convention ratified by Mexico in 1990 (Radwin, 2022).

As researcher Linda Snyder has found, the chance of success in opposing extractive projects increases with a community's ability to remain united. As villagers have reported, mining companies often foster division and hostility within local communities as a way to weaken resisters. Those who side with mining companies may be paid to go door-to-door to promote the company, distribute false information, and spy on, threaten, and at times even attack and kill those who actively oppose mining.

Snyder's observations of an Indigenous community in Guerrero reveal how damaging internal divisions can be. One ejido there, whose water had been contaminated with arsenic from mining activity, joined with other ejidos to address the matter. The mining company responded by making an offer to buy land from the leader of one of the participating ejidos who accepted a deal for well below what the community had been asking, thereby dividing it. After failing in their efforts to meet with state authorities, the coalition set up a blockade at the entrance to the mine to prevent trucks and equipment from entering. These efforts were met with threats and the shooting of one of the activists in 2009. These and other acts of aggression eventually led to the coalition's capitulation to the company's demand for access (Snyder, 2017, pp. 212–13).

As this and many other cases have shown, those who become land defenders do so at considerable risk. Global Witness, an environmental watchdog organization, recently reported that fifty-four environmental activists were killed in Mexico in 2021, making it the most dangerous country in the world for environmental activists (Lopez, 2022, p. 13). Most of these activists were Indigenous. The report also indicates that four mass killings took place that year conveying the message that no one is safe. Such attacks are part of a growing trend in resource rich countries such as Mexico: "A combination of rich natural resources, powerful international companies, violent criminal groups and entrenched government corruption, including in some cases officials who play a role in killings," has turned resource rich areas "hot spots" of violence (Lopez, 2022, p. 13).

A major cause of this violence is a lack of accountability on the part of the Mexican government. Simón Pedro Pérez, a member of the Tzotzil Indigenous Peoples of Chiapas, for example, was gunned down in a local market Chiapas in July 2021, his young son at his side. Just prior to his death, Simón had been preparing to file a complaint regarding extortions and threats from criminal groups working with local governments in an effort to drive Indigenous families from their lands. He was also a member of Las Abejas de Acteal, the Tzotzil pacifist organization which had suffered the massacre in 1997. Pérez had dedicated himself to seeking justice regarding that event (Frabes, 2022). His case is but one of many in which impunity continues to prevail.

Alternatos: An Alternative Vision

For many, the Zapatista Movement has brought dignity to the struggle for land while fostering the dream that "another world is possible". The vision of that world was expressed in the EZLN "declaration for life" in 2021 and translated into eighteen languages. It states, "We make the pains of the earth our own, violence against women; persecution and contempt of those who are different in their affective, emotional, and sexual identity; annihilation of childhood; genocide against native peoples; racism; militarism; exploitation; dispossession; the destruction of nature" (Enlace Zapatista, 2021). As such, the defense of the earth becomes an integral part of the struggle against oppression in its many forms (Maison, 2022).

Both at home and abroad, the Zapatistas continue to promote their political values. Since its founding in 1996, they have backed the National Indigenous Congress as a "home of all indigenous peoples" (Congreso Nacional Indígena, 2017). In 2018, María de Jesús Patricio Martínez, or Marichuy is she is known, ran for president as an affiliate of the EZLN. As the first indigenous woman to run for that office, her goal was less about winning and more about raising the profile of Indigenous peoples throughout Mexico. Zapatistas continue to foster international frienships as well. On a recent trip to Madrid marking the 529th anniversary of Columbus's arrival in the Americas, seven Zapatista delegates met with representatives from 20 different countries as well as a group of Kurdish female combatants who had fought against ISIS (Scott, 2021).

In Chiapas, Zapatistas actively oppose "megaprojects" that damage the earth, disrupt communities, and displace Indigenous and other rural peoples from their land (Scott 2021). Prominent among these is the Mayan Train plan, a 900-mile railway to connect Palenque in Chiapas with other tourist sites in the region at a cost of $10 billion. As with mining projects, Indigenous and other rural peoples are divided on the issue of the train's construction. While some hope that it will bring much needed jobs to the region, others denounce the massive deforestation and despoliation already brought about by the project (Albinson, 2022).

In the Sierra Madre north of San Cristóbal, a youthful staff run Alter Natos, a nonprofit organization founded in 2015 to teach eco-friendly alternatives to the extractive model. Intended as a place "where all worlds fit," Alter Natos seeks to establish a "collective exchange and learning built in harmony, where we promote the search for ecological practices in the face of socio-environmental problems in cities as a different way of relating to the model of forestry, mining, watershed management and of waste (Otros Mundos, n.d.)." The Center consisting of workrooms with vaulted ceilings, library, and dormitory is made largely of clay, bamboo, and various recycled materials. The staff host meetings, tours, and workshops in eco-friendly cooking, plumbing, agriculture, and architecture, as well as Indigenous rights and legal matters (Otros Mundos, n.d.).

Figure 3.2. *View of Alter Natos, sustainable living project of Otros Mundos in Chiapas in 2018.*

In the nineteenth century, Cecil Rhodes gave the world a vision of the sort of riches—gold, silver, and copper—that would build empires and future cities of the world. Then, as now, whatever greatness has come to these mine-fed cities, empires, and personal fortunes has come at a devastating cost, especially to Indigenous Peoples and the land they depend on for their survival. Land defenders in Chiapas offer an alternative vision, one that not only supports the survival of Indigenous and other rural communities, but which offers a safer, more inhabitable environment for the rest of world as well.

Notes

[1] The artist collective: The Electronic disturbance Theater, with Ricardo Dominguez and other collaborators, supported the Internet presence of the Zapatistas movement by creating a network of distribution and information in 1994, that enabled the EZLN to speak to the world without having to pass through the filters of dominant media. Following the Acteal massacre of 1997, EDT organized virtual sits-in and protests, such as *SWARM*, using Floodnet to upload the names of the Acteal victims on Mexican government and Pentagon websites. A 1998 performance titled *Stop the War in Mexico* disrupted the websites of the President of Mexico, of the Pentagon that provided weapons to the Mexican government against the resistance in Chiapas and of the Frankfurt Stock Exchange that aimed to profit from the extraction of uranium and other minerals in Chiapas.

[2] Here the expression "Zapatistas" refers to the movement seeking to advance the rights of Indigenous people in southern Mexico, while the "EZLN" refers to the armed wing of the movement.

[3] These local accounts of rising crime in the region were supported by news coverage in the region, for instance, C. Rodríguez, Narcotráfico y crimen organizado controlan la región del Soconusco: Aranda, Trascender Online Noticias, March 28, 2016, https://trascenderonline.com.mx/narcotrafico-y-crimen-organizado-controlan-la-region-del-soconusco-aranda/

References

Arnaut, J. (2018). Mexican Real Wages Before the Revolution: A Reappraisal. *IberoAmericana*, 47(1), pp.45-62. Available at: https://www.ibero americana.se/articles/10.16993/iberoamericana.421/

Beinart W. (2022). Cecil Rhodes: racial segregation in the Cape Colony and violence in Zimbabwe. *Journal of Southern African Studies*, 48(3), pp. 581–603.

Bessi, R. (2015). El codiciado mineral que amenaza la vida de los pueblos en Chiapas, *Avispa Midia*, 30 noviembre. Available at: https://avispa.org/el-codiciado-mineral-que-amenaza-la-vida-de-los-pueblos-en-chiapas/

General Command of the EZLN (1993). First Declaration from the Lacandon Jungle, Today We Say 'Enough is Enough!' (Ya Basta!). *Modern Latin America*, Web supplement, 8th Edition, Brown University, Center for Digital Scholarship. Available at: https://library.brown.edu/create/modern latinamerica/chapters/chapter-3-mexico/primary-documents-with-accompanying-discussion-questions/document-9-first-declaration-from-the-lacandon-jungle-today-we-say-enough-is-enough-ya-basta-ezln-command-1993/

Calderisi, R. (2021). *Cecil Rhodes and other statues*. Columbus, OH: Gatekeeper Press.

Carr, E. (1964). *The Twenty Years' Crisis 1919-1939: An Introduction to the Study of International Relations*. New York: Harper & Row.

Carrigan, A. (1995.) Chiapas: The First Post-Modern Revolution. *Fletcher Forum of World Affairs*, 19(1), pp. 71-98.

Centro de Informação sobre Empresas e Direitos Humanos (2022). México: El Frente Popular en Defensa del Soconusco (FPDS) sigue luchando para no permitir actividad minera y extractiva en el territorio de Acacoyagua. *Chiapas Paralelo*. Available at: https://www.business-humanrights.org/pt/%C3%BAltimas-not%C3%ADcias/m%C3%A9xico-el-frente-popular-en-defensa-del-soconusco-fpds-sigue-luchando-para-no-permitir-actividad-minera-y-extractiva-en-el-territorio-de-acacoyagua/

Chiapas Paralelo (2020). *Frente Popular en Defensa y la organización que ha detenido la minería en el Soconusco*. Available at:

https://www.chiapasparalelo.com/noticias/chiapas/2020/06/frente-popular-en-defensa-y-la-organizacion-que-ha-detenido-la-mineria-en-el-soconusco/

Chiapas Support Committee. (2018). *Organized peoples of Chiapas against the extractive model.* Available at: https://chiapas-support.org/2018/05/29/organized-peoples-of-chiapas-against-the-extractive-model/

Ciacci, J. (2020). Estamos luchando para sobrevivir. La resistencia a la minería en Acacoyagua, Chiapas. *Circular Tech.* Available at: https://circulartech.apc.org/books/una-guia-sobre-la-economia-circular-de-los-dispositivos-digitales/page/estudio-de-caso-estamos-luchando-para-sobrevivir-la-resistencia-a-la-mineria-en-acacoyagua-chiapas

Ciacci, J. (2021). Extractivism and finite resources. *Association for Progressive Communications.* Available at: https://www.apc.org/en/blog/we-are-struggling-survive-resistance-against-mining-acacoyagua-chiapas

Congreso Nacional Indígena (2017). *Historia del Congreso Nacional Indígena.* Available at: https://www.congresonacionalindigena.org/2017/10/29/historia-del-congreso-nacional-indigena/

Cooper, C. (2021). We are fearful: Indigenous Mexicans dread new military buildup on ancestral land. *The Guardian,* October 4. Available at: https://www.theguardian.com/global-development/2021/oct/04/we-are-fearful-indigenous-tzeltal-mexicans-dread-new-military-buildup-on-their-land-chiapas

Cuentame de México (2020). *Hablantes de lengua indígena.* Available at: https://cuentame.inegi.org.mx/poblacion/lindigena.aspx?tema=P

Dilge, K. (2022). Esperanza Silver Implements Sustainable Development Goals. *Mexico Business News,* May 4. Available at:

https://mexicobusiness.news/mining/news/esperanza-silver-implements-sustainable-development-goals?tag=mining

Easterling, S. (2012). *The Mexican Revolution: a short history 1910-1920,* Chicago: Haymarket Books.

Enlace Zapatista (2021). Primer parte: una declaración… por la vida, 1 enero. Available at: https://enlacezapatista.ezln.org.mx/2021/01/01/primera-parte-una-declaracion-por-la-vida/

Fox, J. (2010). Mexico's indigenous population. *Cultural Survival*, March 26. Available at: https://www.culturalsurvival.org/publications/cultural-survival-quarterly/mexicos-indigenous-population

Frabes, S. (2022). Abejas de Acteal condenan liberación de paramilitares y violencia impune en Chiapas, *Avispa Midia*, 15 agosto. Available at: https://avispa.org/abejas-de-acteal-condenan-liberacion-de-paramilitares-y-violencia-impune-en-chiapas/

Frank, A. (1967). *Capitalism and underdevelopment in Latin America*. New York: Monthly Review Press.

Gobierno de México (2019). *Sistema integral sobre economía minería (SINEM), Dirección General de Desarrollo Minero*. Available at: https://www.sgm.gob.mx/Web/SINEM/mining/mining_companies.html

Grubacic, A. and O'Hearn, D. (2016). *Living at the edges of capitalism: Adventures in Exile and Mutual Aid*. Oakland: University of California Press.

Harvey, N. (1994). Playing with fire: the implications of ejido reform. *Akwe:kon = All of Us: A Journal of Indigenous Reforms*, 20(4).

Hernandez, L. (2010). The San Andrés Accords: Indians and the Soul, *Cultural Survival*, March 26. Available at: https://www.culturalsurvival.org/publications/cultural-survival-quarterly/san-andres-accords-indians-and-soul

Holland, D.L. (2018). "New Extractivism" in Mexico: Hope and Deception. *Journal of Politics in Latin America*, 10(2), pp.123-138. Available at: https://journals.sagepub.com/doi/full/10.1177/1866802X1801000205

In Motion Magazine (2005). Interview with Raymundo Sánchez Barraza: A University Without Shoes. Available at: https://inmotionmagazine.com/global/rsb_int_eng.html

Linares, A., León, V. and Montiel, C. (2022). Mexican environmentalists, Indigenous leaders denounce Mayan Train project. *Noticias Telemundo*, May 11. Available at: https://www.nbcnews.com/news/latino/mexican-environmentalists-indigenous-leaders-denounce-maya-train-proje-rcna28114

Kelly, J. (1994). Article 27 and Mexican Land Reform: The Legacy of Zapata's Dream. *Columbian Human Right Law Review*, 25 Colum. Hum. Rts. L. Rev.

54. Available at: https://scholarship.law.nd.edu/cgi/viewcontent.cgi?article=1693&context=law_facultyscholarship

Lado B. (2019). Resistencia minera en Acacoyagua y Escuintla celebra 4 años. 27 de junio. Available at: https://www.ladobe.com.mx/2019/06/resistencia-minera-en-acacoyagua-y-escuintla-celebra-4-anos/

Lopez, O. (2022). Mexico named deadliest country for green activists. Sept. 29, *New York Times*.

Maison, L.M. (2023). Beyond Western and Indigenous Perspectives on Sustainability: Politicizing Sustainability With the Zapatista Rebellious Education. *Journal of Transformative Education, 21 (1)*. Available at: https://journals.sagepub.com/doi/10.1177/15413446221079595?icid=int.sj-full-text.similar-articles.1

McCafferty, G. (2000). The Chochula Massacre: Factional Histories and Archeology of the Spanish Conquest. In Erwin J. and Hendrickson, M. (Eds.), *The Entangled Past: Integrating History and Archeology*. Archeological Association of the University of Calgary, pp. 347-359. Available at: https://www.academia.edu/210133/2000_The_Cholula_Massacre_Factional_Histories_and_Archaeology_of_the_Spanish_Conquest

Mendoza, J. J. (2015). Costeños se pronuncian contra explotación minera en Chiapas.

Chiapas en Contacto, October 12. Available at: https://www.chiapasencontacto.com/costenos-se-pronuncian-contra-explotacion-minera-en-chiapas-2/

Meredith, M. (2008) *Diamonds, gold, and war: the British and the making of South Africa*, New York: Public Affairs.

Mexico Business News (2013). Hunt for gold and silver in New Spain. October 21. Available at: https://mexicobusiness.news/mining/news/hunt-gold-and-silver-new-spain

Morales, Y. (2022). Por tercera ocasión, Acacoyagua se declara "municipio libre de minería". *Chiapas Paralelo*, 4 septiembre. Available at: https://www.chiapasparalelo.com/noticias/chiapas/2022/09/por-tercera-ocasion-acacoyagua-se-declara-municipio-libre-de-mineria/

Nuijten M., van der Haar, G. (2000). The Zapatistas of Chiapas: Challenges and Contradictions. *European Review of Latin American and Caribbean Studies*, April, pp. 83-90. Available at:

https://research.wur.nl/en/publications/the-zapatistas-of-chiapas-b-challenges-and-contradictions-review-

O'Toole, G. (2017). A constitution corrupted, *NACLA*, 49 (1). Available at: https://nacla.org/news/2017/03/08/constitution-corrupted

Organización Internacional del Trabajo (1996). *Pronunciamiento Conjunto que el Gobierno Federal y el EZLN enviarán a las Instancias de Debate y Decisión Nacional*, 16 enero. Available at: https://www.ilo.org/public/spanish/region/ampro/mdtsanjose/indigenous/pronun.htm

Otros Mundos Chiapas (n. d.). Área del Centro Ecológico "Alter Natos". Available at: https://otrosmundoschiapas.org/area-del-centro-ecologico-alter-natos#:~:text=Alter%20Natos%20es%20un%20espacio,cuencas%20y%20de%20residuos%20que

Osterheld T. (1916). *The history of mining in Mexico and its economic development*. New York: Flanders & Co.

Prebisch, R. (1962). The economic development of Latin America and its principal problems. *Economic Bulletin for Latin America*, VII(1), pp. 1-22. https://repositorio.cepal.org/bitstream/handle/11362/10079/S6200129.pdf?sequence=1

Radwin, M. (2022). Mexico's top court cancels mining concessions near indigenous communities. *Mongabay*, February 18. Available at: https://news.mongabay.com/2022/02/mexicos-top-court-cancels-mining-concessions-near-indigenous-communities/

REMA Chiapas, FPDS, Otros Mundos (2019). ¡Ni en Acacoyagua ni en Brasil! ¡Fuera las empresas mineras del mundo! *Bio Diversidad La*, 31 enero. Available at:

https://www.biodiversidadla.org/Documentos/!Ni-en-Acacoyagua-ni-en-Brasil!-!Fuera-las-empresas-mineras-del-mundo

Rebrii, A. (2020). Zapatistas: lecciones de auto-organización comunitaria. *Democracia Abierta*, 25 Junio. Available at: https://www.opendemocracy.net/es/zapatistas-lecciones-de-auto-organizacion-comunitaria/

Reuters (2022). Mexican mining industry expects investments to rise 15% in 2022. August 23. Available at: https://www.mining.com/web/mexicos-mining-industry-expects-investments-to-rise-15-2-in-2022/

Rodríguez, C. (2016). Narcotráfico y crimen organizado controlan la región del Soconusco: Aranda. *Trascender Online Noticias*, 28 marzo. Available at: https://trascenderonline.com.mx/narcotrafico-y-crimen-organizado-controlan-la-region-del-soconusco-aranda/

Rostow, W. W. (1960). *The stages of economic growth: a non-communist manifesto.* Cambridge: Cambridge University Press.

Sachs, J. (2005) *The End of Poverty: Economic Possibilities for Our Time.* London: The Penguin Press.

Salvemos, C. (2011). *El oro de los tontos.* Available at: https://www.youtube.com/watch?v=aj3C2RlV6iw

Scott, F. (2021). "We were never conquered": Mexico's Indigenous Zapatista movement visits Madrid. *Madrid No Frills,* Oct.11. Available at: https://madridnofrills.com/the-history-of-the-anti-colonialism-zapatista-movement/

Snyder, L. (2017). *Struggles for justice in Canada and Mexico.* Waterloo, ON: Wilfrid Laurier University Press.

Soberanes, R. (2017). Comunidades se oponen a 21 proyectos mineros en la Sierra Madre de México. *Mongabay,* Oct. 20. Available at: https://es.mongabay.com/2017/10/no-la-mineria-la-lucha-conservar-la-sierra-madre-mexico/

Stoltenborg D. and Boelens, R. (2016). Disputes over land and water rights in gold mining: the case of Cerro de San Pedro, Mexico. *Water International,* 41(3), pp. 447–467. Available at: www.tandfonline.com/doi/pdf/10.1080/02508060.2016.1143202

Stone, H. (1978). Mexican agrarian rights: who do they benefit? *California Western International Law Journal,* 8(1), pp.130-170. Available at: https://scholarlycommons.law.cwsl.edu/cgi/viewcontent.cgi?article=1800&context=cwilj

Tetreault, D. (2014). Free-market mining in Mexico. *Critical Sociology,* 42(4-5), pp. 1–17. Available at: http://journals.sagepub.com/doi/abs/10.1177/0896920514540188

Tetreault, D. (2015). Social environmental mining conflicts in Mexico. *Latin American Perspectives,* 42(5), pp.48-66.

Tilly, C. and Kelly, M. (2006). Mexico: the Zapatistas' new fight. *Against the Current*, (123), July-August.

Toussaint, E. (2017). Mexico proved that debt can be repudiated. *CADTM newsletter*, March 24. Available at: http://www.cadtm.org/Mexico-proved-that-debt-can-be#:~:text=The%20Porfiriato%20lasted%20until%20the,and%20internal%20public%20debt%20doubled

Varzi, C. (2020). 23 Years of Impunity for Perpetrators of Acteal Massacre. *NACLA*, 52(4). Available at: https://nacla.org/news/2020/12/21/23-years-impunity-acteal-massacre

Wager, S.J. and Schultz, D.E. (1995). Civil-military relations in Mexico: The Zapatista revolt and its implications. *Journal of International Studies and World Politics*. 37(1), pp. 1-42.

Warman, A. (2003). La reforma agrarian mexicana: una visión de largo plazo, *Reforma agraria, colonización y cooperativas*, 2, pp.85-94. Available at: https://www.fao.org/3/j0415t/j0415t09.htm#TopOfPage

WorldAtlas (n. d.). The poorest states in Mexico. Available at: https://www.worldatlas.com/articles/the-poorest-states-in-mexico.html

Chapter 4

The River as a Common Good: Carolina Caycedo's Cosmotarrayas

Carolina Caycedo and Jeffrey De Blois

This essay combines excerpts of a multimedia essay[1]that appeared in conjunction with the exhibition "Carolina Caycedo: Cosmotarrayas" at the Institute of Contemporary Art, Boston, 2020.

Be Dammed

Indigenous cosmogonies throughout the Americas understand water to be the sacred essence of life. All bodies of water are interconnected, rivers are the veins of the planet, drawing together different communities and ecosystems. Rivers, like all bodies of water, are living entities, and as such, are subjects rather than objects. As the effects of climate change are more widely acknowledged, many have turned toward such indigenous knowledge in the effort to foster healthier relationships to the environment and non-human life. In the twenty-first century, water is already one of the most disputed natural resources. In the United States, water protectors protested against the Dakota Access Pipeline at Standing Rock Indian Reservation, united by the slogan "Water is life." The water protectors active in the Dakotas stood in solidarity with individuals and groups across the world demanding fair access to water and those experiencing water insecurity. Around the same time, in several Latin American countries (following a historic ruling in New Zealand) national governments began to grant rivers and other natural bodies environmental personhood, "a nascent notion of designating parts of nature as legal persons entitled to independent regard and considerations" (Gordon, 2017). Ecuador's 2008 constitution, for example, proclaims the rights of nature "to exist, persist, maintain, and regenerate its vital cycles"[2]. In 2010, Bolivia passed the "Law of the Rights of Mother Earth," designating Mother Earth "a collective subject of public interest" (Law 071, 2010). After illegal mining in the Atrato River Basin in northwestern Colombia produced devastating natural and humanitarian impacts, the Colombian Constitutional Court declared in 2016 that the Atrato River possesses rights to "protection, conservation, maintenance, and restoration" (CELDF, 2017).

This historic ruling came only a few years after a "Master Plan" for the development of the Magdalena River was signed by the Colombian and Chinese governments in 2014. The Magdalena is Colombia's largest and most important river, stretching nearly 1,000 miles from the mountainous region in the southwest near Ecuador to the Caribbean Sea in the north[3]. Along with other provisions, the "Master Plan" called for the construction of up to fifteen hydroelectric dams on the Magdalena, in addition to one other dam constructed in 1987 and another under construction since 2011. Activists against the construction of new hydroelectric dams on the Magdalena point out that they are not necessary, as Colombia already produces enough energy to meet domestic demands from hydroelectric power generated from other major rivers, plus surplus power it exports to Venezuela, Panama, and Ecuador (Baumhardt, 2015). Nevertheless, in the wake of fifty years of conflict in Colombia, foreign development raced to extract resources from the Magdalena (Baumhardt, 2015).

It is against this backdrop that the Los Angeles-based, Colombian artist Carolina Caycedo (b. 1978, London) initiated her ongoing project *Be Dammed* in 2012, to examine the wide-reaching impacts of dams built along waterways, particularly those in Latin American countries. The aims of this project are to explore, as she says, "how infrastructures of an extractivist nature... affect social bodies and bodies of water" (Bonhomme, 2017). Caycedo was raised in the greater Magdalena River Basin and frequently returns to Colombia where she is an active member of Rios Vivos Colombia Social Movement and is involved on a national and international level in movements of solidarity in the struggle for water, land, and life. In 2012, more than 250 large hydroelectric dams were planned or under construction by transnational corporations across Latin America, signifying the transition of bodies of water from public entities to privatized resources. Simultaneously, the United States is the leading country in dam removal, allowing for the restoration of river ecosystems. Around 900 dams were removed in the US between 1990 and 2015. These cursory details demonstrate the asymmetrical power relations and disjunctive modes of governance at work between countries and multinational corporations in the so-called Global North and those in the so-called Global South, where forms of colonial violence and oppression are still in operation.

Be Dammed takes several different forms—from workshops and collective actions that Caycedo refers to as "geochoreographies," to a series of large-scale photographic Water Portraits, to an accordion-style artist's book called *Serpent River Book* (2017). Many of the projects in *Be Dammed* incorporate indigenous forms of knowledge. Each project grows organically out of what Caycedo refers to as "spiritual fieldwork" and her intimate and ongoing relationships with individuals and groups in different riverine communities adversely affected by the privatization of waterways. For her research, Caycedo "blend[s] into the local and regional agendas, keeping aesthetic intention and art production in the background" (Lozano, Caycedo, 2018). She then "proceeds to visit, record, collect objects, and create friendships with people who live and work there, sharing the intimacy of homes, places of work, and spaces dedicated to direct action, learning, and collective celebration" (Lozano, Caycedo, 2018). Many of the individuals she met doing fieldwork are recurring figures in her work, such as a fisherwoman, day laborer, and community organizer from La Jagua, Colombia named Zoila Ninco or Don Élio Alves da Silva, a fisherman displaced by the Belo Monte dam on the Xingu River in Brazil. Whereas dams in many parts of the world have been "naturalized" in some ways and are commonly viewed as benign, the incursion of multinational corporations to build dams in Latin America has been met with protest. Kogui indigenous spiritual leader Mamo Pedro Juan, who passed away in 2017, described a dam as severing the connection between bodies of water and communities, "like a knot in the veins, or, still worse, a knot in the anus" (Caycedo, 2017). The opposition to dams by activists, farmers, fisherfolk, and others has been persistent, even while Colombia is consistently found to be in the top three nations globally for murder rates of environmental defenders.

Be Dammed is directed at constructing what Caycedo refers to as a "popular historical environmental memory" that might produce embodied forms of everyday knowledge and counternarratives to the dominant ideologies of multinational corporations and destructive tendencies of extractive capitalism. In the context of environmental personhood, and by telling the stories of those affected, Caycedo's work proves vital in the documentation and aesthetic mediation of urgent issues related to a "collective subject of public interest" and in its multivalent aesthetic strategies intended to resist and counteract toxic ideologies and repetitious cycles of dispossession and violence.

This essay traces the genealogy of the cast fishing net, or *atarraya* in Spanish, in Caycedo's practice—a resonant object that appears first in her videos and performances before she produces a series of hanging sculptures called Cosmotarrayas. One component of *Be Dammed*, the Cosmotarrayas are assembled with handmade fishing nets and other objects collected during her field research in riverine communities. The rich story of the cast fishing net in Caycedo's work, one that interweaves specific people, rivers, traditions, and cultures, is recounted here through text, images, documentation, and storytelling. The Cosmotarrayas are deeply symbolic objects, and reinforce Caycedo's tenet that, "to cast a fishing net affirms the river as a common good" (Caycedo, 2016).

Spaniards Named Her Magdalena, But Natives Call Her Yuma

The cast fishing net first appeared in Caycedo's work in the two-channel video essay *Spaniards Named Her Magdalena, But Natives Call Her Yuma* (2013). Made during a DAAD residency in Berlin, *Spaniards Named Her Magdalena, But Natives Call Her Yuma* juxtaposes footage of dams and reservoirs in Germany with Caycedo whispering a bilingual narrative about her relationship to rivers and how dams become "naturalized" over time. Renamed Magdalena in 1501 by the Spanish during their colonization of the Americas, the river was first called Yuma, which means "land of friends," by the Muisca confederation several centuries earlier, a name still in use by local inhabitants. During her field research, Caycedo interviewed a broad range of individuals either effected by or involved with the development of the controversial El Quimbo Hydroelectric Project in the Huila Department. El Quimbo was the first hydroelectric power project to be constructed in Colombia by a multinational corporation, the Spanish hydroelectric utility company Endesa via its Colombian subsidiary Emgesa. This massive concrete faced rock-fill dam radically diminished fish populations, compromising the economic autonomy and food sovereignty of riverine communities who depend on the Magdalena for their way of life.[4]

In addition to an activist, an environmentalist, an opposition leader, a professor, a shaman, and the dam's engineer, Caycedo interviewed local fisherwoman Zoila Ninco. Ninco was interviewed for a brochure produced for Caycedo's 2013 exhibition *The Headlong Stream is Termed Violent, but the Riverbed Hemming Is Termed Violence by No One* at DAAD Gallery in Berlin where she is identified anonymously as "Fisherwoman." In *Spaniards Named Her Magdalena, But Natives Call Her Yuma*, Zoila Ninco stands on the banks of a rushing river on the right, next to footage of stagnant water in a German reservoir on the left. Later in the video, Zoila Ninco appears again casting her fishing net, alongside footage of Caycedo herself trying to scale the wall of a German dam. Zoila Ninco features in other videos, drawings, and performative actions, and her family even provided many of the cast fishing nets used for the Cosmotarrayas, including the nets for the first three sculptures produced in 2016.

Land of Friends

Caycedo continued her critical focus on the Magdalena River, colonial histories, and the El Quimbo Hydroelectric Project in the video essay *Land of Friends* (2014), which expands upon the artist's multiperspectival approach in novel ways.[5] Whereas *Spaniards Named Her Magdalena, But Natives Call Her Yuma* features a bilingual first-person narrative whispered by Caycedo herself, *Land of Friends* features the voices of many of the individuals interviewed in the fieldwork for the earlier video as testaments.

Activists, politicians, fisherfolk, engineers, farmers, and Caycedo's own family feature, as do the "submerged" perspectives of non-human agents—the baby armored catfish, the firefly, and the leaf cutter ant (Gomez-Barris, 2017, 95). The central protagonist of *Land of Friends* is the Magdalena itself. Through various methods, like the use of satellite imagery, long shots of the river's movement and surrounding environment, its sound, and its voice, Caycedo portrays the river as a living being. Zoila Ninco appears on the banks of the Cuacua, or Suaza, River, with her net, repeatedly casting and pulling back in as Caycedo describes who she is, before unspoken text captions compare her to Cacica Gaitana, a political and religious leader that in 1539 unified several indigenous tribes to defend the Huila territory from the Spanish conquest. It is in witnessing Zoila Ninco perform this repetitive gesture, and in thinking about the river's history, that Caycedo becomes fascinated with the cast fishing net as an affective object. For Caycedo, the material qualities of the fishing net—they are porous, malleable, handmade, and embody ancestral knowledge—offer a potent counterpoint to the brute-force infrastructure of dams, which disrupt the natural flow of rivers, dispossess people of their homes, and threaten their way of life.

Cosmotarrayas

The first Cosmotarrayas were portraits of three disputed rivers—the Yuma in Colombia, the Yaqui in the state of Sonora in northwestern Mexico, and the Elwha on the Olympic Peninsula in Washington—embodied first as characters in the play *One Body of Water* (2015). At the time, the Yuma, Yaqui, and Elwha symbolized rivers at different stages of privatization or restoration in the Americas. While the Yuma in Colombia was in the initial stages of privatization with two hydroelectric dams completed and a further fifteen dams projected according to the "Master Plan," the Yaqui River had already been completely privatized with three dams and several irrigation channels causing the river's mouth to dry up, leaving eight traditional Yaqui villages without water. In stark contrast, the Elwha River in the Pacific Northwest had undergone the largest dam removal in US history thanks to the Elwha River Ecosystem and Fisheries Restoration Act of 1992 that initiated the removal of the Elwha Dam and the Glines Canyon Dam. The removal of these dams allowed for the salmon to spawn after 100 years of not being able to swim upstream when the river's natural flow from its headwaters in the Olympic Mountains to the Strait of Juan de Fuca was restored.

One Body of Water was developed in response to the site at Bowtie Project in Los Angeles, with Caycedo and two collaborators giving voice to the rivers around a fire, employing indigenous oral traditions to rupture dominant ontological positions that structure the way we think about nature and how we interact with the environment. During the play, the performers' painted faces were inspired by traditional indigenous masks from these rivers, and each wore a different poncho and carried instruments and objects that signified the rivers they anthropomorphized. The Cosmotarrayas *Yuma, Yaqui,* and *Elwha* (all 2015) assembled these various objects and others within fishing nets supplied by the family of Zoila Ninco.

Yuma, Yaqui, and *Elwha* were first presented in the 2016 group exhibition *Entre Caníbales* (Among Cannibals) at the gallery Instituto de Visión in Bogota. The exhibition featured artists with a shared interest in "examining the tools, consequences and catastrophes of colonization processes on human communities, nature, and food," according to Beatriz López, the exhibition's curator (Lopez, 2016). Several of Caycedo's works were featured in the exhibition, including *Land of Friends*, the performance script of *One Body of Water*, and her hand-bound, accordion-style illustrated River Books for the Yuma, Yaqui, and Elwha Rivers, all of which introduced the Cosmotarrayas in the broader context of her expanded practice.

Following this critical shift from employing the cast fishing net as a resonant performance object to its incorporation as a structuring material and concept in sculptural assemblage, Caycedo would go on to create a rich and varied series of hanging sculptures called Cosmotarrayas. Each Cosmotarraya is assembled with handmade artisanal fishing nets, often dyed by the artist, and various symbolic objects collected during her field research. Cosmotarraya combines the words "cosmos" and *"atarraya"* (Spanish for net) to form a compound that conveys the centrality of the net in the life of those who fish. Each Cosmotarraya is linked to specific people, rivers, traditions, and cultures. Likewise, each net is connected to an individual body, woven by hand with a needle made to the thickness of an individual fisherperson's fingers, as demonstrated by Antonio Chavarro weaving a net in the social documentary *Huila's Bleeding* (Caycedo, Aguas, 2015). For Caycedo, the Cosmotarrayas are talismanic objects that cast visual spells: they embody the continued resistance to corporations and governments seeking to control the flow of water, create visual narratives that counter the supposed neutrality of dams, and act as a critical bridge between Caycedo's community-based work and her studio practice. The Cosmotarrayas cast a net by other means to different ends.

Hunger as a Teacher/El Hambre Como Maestra

In 2017, for a solo exhibition at the Los Angeles gallery Commonwealth and Council entitled *Hunger as a Teacher/El Hambre Como Maestra*, Caycedo continued to develop new material and conceptual approaches to the Cosmotarrayas, summoning the notion of dams and brute-force infrastructures at human scale and in specific relations to the body. As she describes it, "I approached the making of these nets as counter narratives to visual regimes of containment" (Caycedo, 2019). The title of the exhibition was drawn from a conversation Caycedo had with Brazilian fisherwoman Raimunda da Silva. When asked who taught her how to fish, Raimunda answered, "Hunger taught me to fish."

While the earliest Cosmotarrayas are like object-based portraits of rivers, and though many of the sculptures made in Brazil referenced individuals and communities she met there, the group of sculptures made for *Hunger as a Teacher* developed an aspect only briefly introduced in Brazil: Cosmotarrayas as talismanic, ritual objects that conjure or give offerings to spiritual or folkloric entities. In these objects, Caycedo adopts gestures associated with witchcraft—such as binding or mixing potions—as aesthetic strategies associated with healing. Sculptures such as *Big Woman/Mujer grande* or *She Came from the Deep/Ella vino de lo profundal* (both 2017) summon the feminine force of nature. As described by the artist Candice Lin, *Big Woman* appears to be, "a witchlike, folk female figure with a painted wooden mask for a face," who "seems to hover protectively, omen-like, promising plentitude yet, perhaps, righteous retribution if one transgresses upon the bounty of her body" (Lin, 2018, pp.24-5). For another sculpture, *To Drive Away Whiteness/Para alejar la blancura* (2017), Caycedo concocted potions in glass bottles using ingredients such as chili peppers, achiote, dried kelp seeds, hibiscus, and ginseng mixed with water from the Pacific Ocean, the Los Angeles River, and the Colorado River (which is the source of Los Angeles tap water). These potions are imagined to be magical elixirs to counter whiteness, with whiteness here being synonymous with the destructive Western world view that sees nature as a resource to be mined.[6]

Caycedo also considered her own body in relationship to regimes of containment, and for *Undammed/Desbloqueada* (2017) removed the contraceptive intrauterine device that had been "damming" her body, placing it within a conical net, just above a Navajo sandstone resting on a metal gold pan. Through this gesture, Caycedo had personally rebuked Western forms of birth control, deciding to turn instead to contraceptive methods attuned to the body's natural rhythms. In contrast to previous installations that referenced fishing nets hanging to dry, the Cosmotarrayas in *Hunger as a Teacher* were stretched out and displayed according to their circularity, to accentuate the high level of sophisticated craftsmanship and labor that goes into weaving each net, and to express their full scale.

Made in L.A.

Several of these sculptures, as well as more recent Cosmotarrayas, were suspended from the second level above the Hammer Museum's outdoor courtyard as part of Caycedo's inclusion in the 2018 edition of the recurring biannual exhibition *Made in L.A.* This installation recalled the architectural intervention at the Bienal de São Paulo, but here the sculptures were outside, making them prone to ambient environmental conditions, to explore kinetic movement brought about by the wind, and the unique interactions of light and shadows produced by the sun moving across the sky. Caycedo experimented here by combining several nets together to form a single, large-scale work, like *Milk, Sap, Currents, Blood and Fire* or *Curative Mouth* (both 2018). In *Milk, Sap, Currents, Blood and Fire* the colorful nets were embroidered with symbols like the map of a river's currents or a tree with its subterranean network of roots. Another Cosmotarraya was made in reverence to Ósun, the Yoruba river spirit of water, pleasure, fertility, and sexuality. Like *Big Woman* and *She Came from the Deep*, Ósun is an offering to a spiritual entity whose hand-painted image on a steel pot lid gives the Cosmotarraya its conical shape. There, Ósun is a mermaid-like figure holding a forked tail made up by two bodies of water. She is accompanied by an incantatory text in Spanish that reads, "rios vivos, pueblos libres": living rivers, free people.

A Cobra Grande

Caycedo realized her largest Cosmotarraya, *A Cobra Grande* (2019), for the exhibition *The Green Goddess* staged in 2019 in Lille, France as part of a city-wide thematic initiative of exhibitions and programs called *Eldorado. A Cobra Grande*, a large-scale sculpture comprised of several interconnected nets dyed fluorescent shades of green, yellow, and fuchsia takes the form of a snake, inspired by Amazonian folklore about a "big snake" that inhabits the river's depths and is responsible for the formation of new waterways, or *igarapés*, as it meanders around the forest. The big snake is said to be responsible for the mudslides and other mass wasting events when angered by human foolishness. A compelling narrative often instrumental in helping to explain otherwise inexplicable environmental phenomena, *Cobra Grande* is also "a central Amazonian metaphor," according to Nicholas C. Kawa, "that reminds us that our surroundings are in constant flux, and that humans are not the only ones responsible for this ongoing transformation"(Kawa, 2015) This figure of the big snake, imaginary or not, is a reminder of the continuous presence of non-human forces at work in the world around us.

Flying Massachusetts

Caycedo's solo exhibition at the Institute of Contemporary Art/Boston was the largest gathering of her Cosmotarrayas to date, for which she produced *Flying Massachusetts* (2020), her first sculpture imagined to be the physical expression of a land acknowledgment. According to Caycedo, "It is an acknowledgment to Massachusett, the sacred Great Blue Hill that overlooks the Boston Harbor and that hosts my work; and to the people who traditionally inhabited the Greater Boston area, and today continue to live and relate to the lands and waters as the Massachusett Tribe at Ponkapoag, and the Natick Massachusett-Nipmuc" (Caycedo, 2019). Like other of the most recent Cosmotarrayas that map the spiritual possibility of devotional sculptures, *Flying Massachusetts* is an object grounded by the reparative gesture of the land acknowledgement. Conjuring non-human perspectives is central to Caycedo's most recent work and many of her Cosmotarrayas are informed primarily by indigenous and Native American epistemologies and philosophies that recognize water, land, animals, plants, and minerals as living subjects with agency, capable of transforming our surroundings. Like all of her work, *Flying Massachusetts* is rooted in a profoundly ethical engagement with the world—towards humans and non-humans alike—in service of imagining counter visual narratives aimed pointedly at breaking toxic cycles of development, dispossession, and violence.

Notes

[1] Available at: https://www.icaboston.org/publications/river-common-good-carolina-caycedos-cosmotarrayas

[2] The rights of nature are described under Title II, Chapter 7 of the Constitution of the Republic of Ecuador, last updated January 31, 2011. Available at: http://pdba.georgetown.edu/Constitutions/Ecuador/english 08.html

[3] The Magdalena supplies water to 80% of Colombia's population and 86% of the country's gross domestic product is generated in its basin.

[4] In 2000, the World Commission on Dams published a report entitled "Dams and Development," which concludes that large dams (15 meters or higher) are not sustainable, due to their social, environmental, and economic impacts to local ecosystems and communities. El Quimbo is 151 meters high.

---

[5] *Land of Friends* was Caycedo's thesis project for the completion of her M.F.A. at the University of Southern California's Roski School of Art and Design. The video is available at: https://vimeo.com/94685623

[6] For the ICA/Boston presentation of *To Drive Away Whiteness: Para alejar la blancura,* the three water sources for the potions are the Atlantic Ocean, the Charles River, and the Quabbin Reservoir.

References

Baumhardt, A. (2015). The Magdalena and the "Master Plan", *Los Angeles Review of Books*, March 7, 2015. Available at: https://lareviewofbooks.org/article/magdalena-master-plan/

Bolivia's Plurinational Legislative Assembly (2010). *Ley de Derechos de la Madre Tierra (Law of the Rights of Mother Earth)*. Chapter II, Article 5.

Bonhomme, H. (2017). Carolina Caycedo Finds the Beauty in Continuity, *Creative Capital*.Available at: https://creative-capital.org/2017/10/27/carolina-caycedo-finds-the-beauty-in-continuity/

Caycedo, C. (2016). *Cosmotarrayas*. Available at: http://carolinacaycedo.com/cosmotarrayas-comotarrafas-series-2016

Caycedo, C. (2019). *The River as a Common Good: Carolina Caycedo's Cosmotarrayas*. Preparatory notes.

Caycedo, C. and Aguas, E. (2015). *Huila's Bleeding*. Available at: https://www.youtube.com/watch?v=aa66TqNE12Q and: https://creativetimereports.org/2015/03/17/activists-fight-privatization-colombias-magdalena-river/

Commonwealth and Council. (2017). *Carolina Caycedo: Hunger as a Teacher/ El Hambre Como Maestra*. Press release. Available at: http://commonwealthandcouncil.com/exhibitions/hunger-as-teacher-el-hambre-como-maestra/press

Community Environmental Legal Defense Fund (2017). *Colombia Constitutional Court Finds Atrato River Possesses Rights*, May 4. Available at: https://celdf.org/2017/05/press-release-colombia-constitutional-court-finds-atrato-river-possesses-rights/

Gómez-Barris, M. (2017). A Fish-Eye Episteme: Seeing Below the River's Colonization. In: *The Extractive Zone: Social Ecologies and Decolonial Perspectives*, p. 95. Durham, N.C.: Duke University Press.

Gordon, G. (2017). *Environmental Personhood*. University of Pennsylvania, The Wharton School. Available at: http://dx.doi.org/10.2139/ssrn.2935007

Kawa, N.C. (2015). *Cobra Grande: An Amazonian Vision of Human Environmental Relations*.Engagement: Anthropology and Environment Society. Available at: https://aesengagement.wordpress.com/2015/11/10/cobra-grande-an-amazonian-vision-of-human-environmental-relations/

Lin, C. (Summer 2018). Licking the Wound: Three Works from Pacific Standard Time: LA/LA, *X-TRA*, 20(4), pp.24-5. Available at: https://www.x-traonline.org/article/licking-the-wound-three-works-from-pacific-standard-time-la-la

López,B. (2019). Entre Caníbales. *Instituto de Visión*, April 16-June 19, Bogota. Available at: http://institutodevision.com/exposiciones/entre-canibales/

Lozano, C. Caycedo, C. (2018). Nunca fuimos modernas: Catalina Lozano en

Conversación con Carolina Caycedo. *Terremoto 12*. Available at: https://terremoto.mx/article/nunca-fuimos-modernas/

Cosmotarrayas: Carolina Caycedo

Figure 4.1
Big Woman/Mujer Grande, 2017
Wooden mask, handmade fishing net, handmade wool hammock, nylon fishing net, fabric, dry cattails, dry plantain stem fibers, vine, and rope.
Approximately 85 x 26 x 62 in (215.9 x 66.04 x 157.47 cm)
Collection Museum, Lodz, Poland

Figure 4.2
Land of Friends / Tierra de Amigos, 2014
One channel HD Video. Sound and Color. 38 minutes 10 seconds

Figure 4.3

Made in L.A., Installation view, June 3–September 2, 2018

Hammer Museum, Los Angeles.

Photo: Brian Forrest

Figure 4.4
Flying Massachusett, 2020
Hand-dyed artisanal fishing net, artisanal fishing trap with floaters, hand-dyed artisanal hammock, traditional textile from the Zamboanga Peninsula in the Philippines, hand-carved wooden gold pan, hand-carved wooden boat, white shell, wooden needle, pebbles and stones collected from Boston Harbor, Blue Hills Reservation, and Neponset River, tin jingle cones, cotton thread, and paradors
93 x 166 x 23 inches (236.2 x 421.6 x 58.4 cm.)
Collection of Benedicta M. Badia
Installation view, Carolina Caycedo: Cosmotarrayas, January 20-September, 2020, ICA Boston.

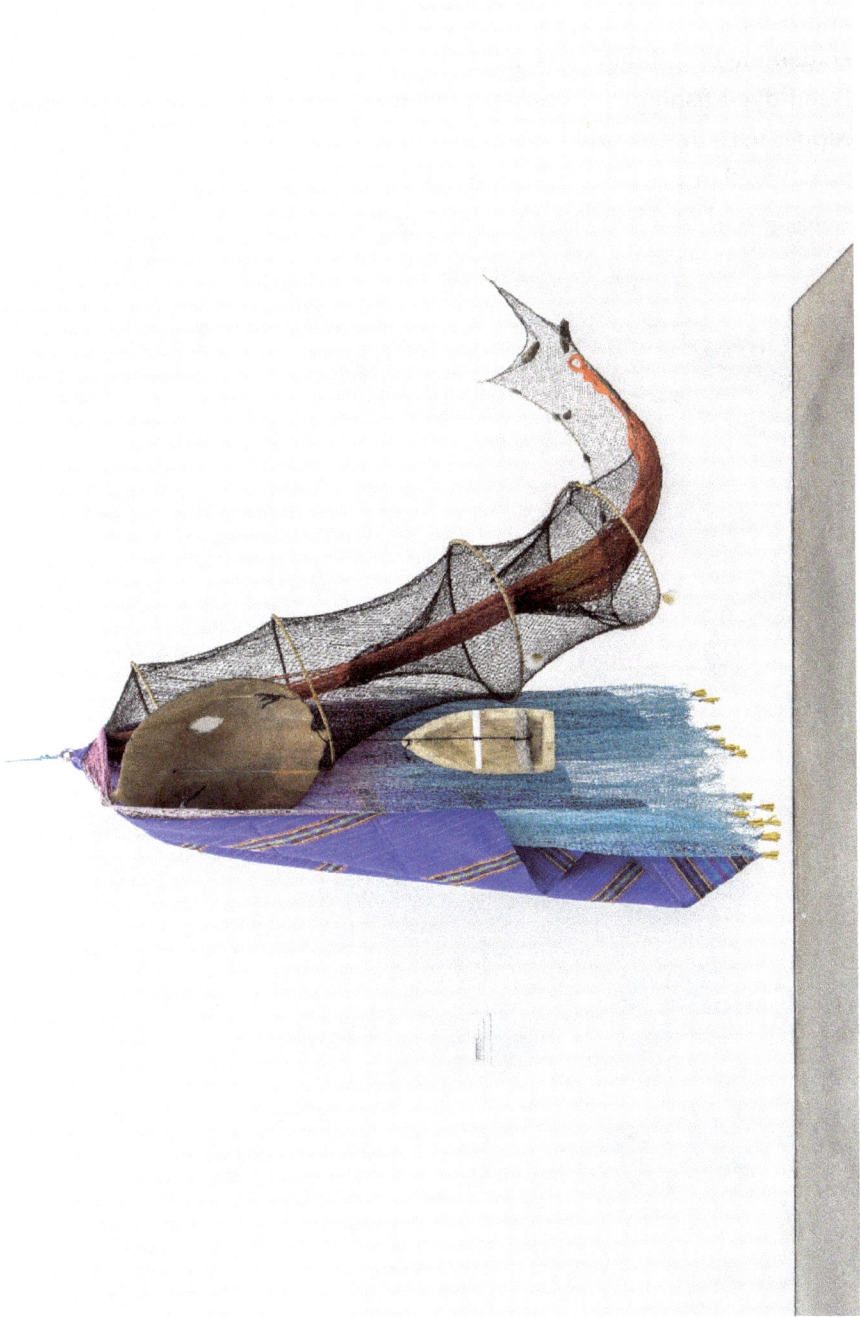

Figure 4.5
Undammed/Desbloqueada, 2017
Hand-dyed fishing net, lead weights, metal gold pan, Navajo sandstone, copper IUD, thread, and rope.
64 x 19 x 19 inches (162.5 x 48 x 48 cm).
Collection of Ann Soh Woods, Los Angeles

Part 2

Resisting Extractivism: The Centrality of
Indigenous Voices

Chapter 5
Mni Wiconi –
Water is [more than] Life

Edward Valandra

This essay was first published as a chapter in Standing with
Standing Rock: Voices from the #NODAPL Movement,
by University of Minnesota Press, 2019.[1]

Why Stand with Water

"MNI WICONI!" The Water Protectors' call circled the world in late 2016, and for many people, the call continues to resonate. This collective cry captures more than the human experience with water. Yes, we understand that all water-dependent life perishes without healthy water. We all have experienced thirst that only water quenches. We know almost all flora and fauna struggle and die without water; the World Wide Web alerts us immediately to drought-stricken areas. And, for the better part of history, Western-based societies have used water as a conventional utility in the service of humankind, for example, hydroelectric power.

So when both the Dakota Access Pipeline (DAPL) and Keystone XL Pipeline (KXL) invaded Oceti Sakowin Oyate homeland, endangering the water for millions, we challenged the Western assumptions behind these projects, especially the DAPL pipeline—# NoDAPL. Our initial challenge seemed orthodox enough. Prior to our Hunkpapa Titunwan relatives establishing the Sacred Stone Camp, they, through colonizer political channels such as the National Environmental Policy Act and National Historic Preservation Act, informed Energy Transfer Partners (ETP), DAPL supporters such as North Dakota, the US Army Corps of Engineers, and the Interior Department of their opposition to DAPL.

Going deeper, though, our people also challenged the fundamental mindset behind the projects—the Western metaphysics driving modern development. For example, in another essay, I challenged the "positive development" ideology that DAPL's backers used to justify it and to overlook its dangers:

This $3.7 billion, 1,172-mile pipeline will transport between 470,000 to 570,000 barrels of oil per day from the Bakken/Three Forks formations in North Dakota for domestic consumption: "Its goal is to relieve transportation strains on rail for crude transportation and safely transport U. crude oil to US markets via pipeline to further the goal of energy independence."

The stated goal, of course, appeals to mainstream thinking, US nationalism, and economics: Americans will be less dependent on foreign oil, while simultaneously growing their economy. For example, a few of the windfalls that Energy Transfer Partners (ETP) attributes to DAPL include labor income ($1.9 billion); right-of-way payment of landowners ($190 million); local use, gross receipts, and lodging taxes during construction ($10 million); and state individual income tax revenue ($28million). These monetary figures reveal the Western assumption that development is inherently beneficial and everyone wins (Valandra, 2016).

When both the Sacred Stone and Spirit Camps were established; when the Oceti Sakowin Oyate reignited its ancient, sovereign fire; and when thousands of people from all over the world either descended on the Spirit Camp as Water Protectors or tracked it on social media: water-is-life surged through consciousness, not solely as an environmental cause célèbre but as Mni Oyate (Water Nation). Development's excesses (pollution, contamination, damming) show the West's blindness to who water is, and this blindness dictated the United States' response. Unable or unwilling to distinguish Water Protectors from conventional protesters, and unable or unwilling to differentiate between water as a quantifiable utility, for example, acre-foot, and water as a rights holder, that is, as possessing personhood, the United States defaulted to its standard post-9/11 policy, a militarized police reaction. Corporate-influenced governments showed zero tolerance for ways of knowing that question a Westernized lifestyle (Kluger, 1976).

This essay first discusses Indigenous sovereignty. Sovereignty is germane to the DAPL controversy because the struggle is fundamentally a culture or paradigm war. Without first examining sovereignty, the uninitiated get lost in the political complexities and fail to name what North Dakota's and the United States' actions really represent: differences in world view and colonization. Second, I show how the narrowness of the Western concept of legal standing impacts nonhumans and their status as rights holders. Third, since the first two discussions are about comparative world views, I formulate how whites apply their Otherizing of nonwhites to the Natural World as well. Fourth, I discuss Mni Wiconi from our perspective and how it informs our resistance to DAPL. Finally, I dissect how colonization works hand in glove with corporate development to marginalize our voices, using Western procedures to do it. Here ETP's narrative about its decision-making process for DAPL is one-sided, hence self-serving. The corporate process favored non-Indigenous protocols, and the subsequent narratives from such protocols worked against both Indigenous communities and the Natural World. Naming this colonized process helps us understand DAPL's political strategy: to present ETP as a good neighbor when it is not. Moreover, the state's militarized police and the corporate security reaction to the Oceti Sakowin Oyate and its allies exposed ETP's Mni Wiconi good neighbor rhetoric as nothing more than a public relations lie. It is about the genocide of a people whom we call the Mni Oyate.

Indigenous Sovereignty Is All About My Relatives

Why did we, the Oceti Sakowin Oyate, take the stand we did? This question opens a discussion on a core issue rarely heard in Western circles: Indigenous sovereignty and where it originates. The Oceti Sakowin Oyate predates modern states, and our sovereignty flows directly from our origin story: how we came to be and the primary responsibility given us. By framing our existence in relationship with the Natural World, our story counters the Western assumption that the Natural World's sole purpose is to serve humankind—the tenet of today's Anthropocene era. Our oral stories remind us that such a mindset, sovereignty without the Natural World's inclusion, leads to Earthwide extinction as a wholly human-induced catastrophe (Kolbert,2014). In fact, today's mass extinction replays one of our oral history stories, "The Great Race." Sometime in ages past, the Oceti Sakowin Oyate did not live as good relatives with the Natural World. As a result, our right to exist lay in nonhumans' hands. The Water Protectors' actions within our homeland prove that the Oceti Sakowin Oyate learned our lesson from the race. We must follow our original instructions: *to be a good relative*. Since water is our relative, we protect all our relatives from harm. Resisting DAPL was our response to a relative's call for help; answering that call was and remains our responsibility.

Compared to Indigenous sovereignty, Western sovereignty is a construct of human thinking, unconnected to the Natural World on which humans depend, hence a fiction. Western sovereignty's ultimate expression is the modern state, the basic characteristics of which include a bounded territory, absolute jurisdiction within its exterior boundaries, a self-recognized legitimacy, and a political architecture that provides for national expression (e.g., war declaration, a national capital, chosen leaders, bureaucracy, social contract or compact). Since the mid-twentieth century, the United Nations (UN) has become the organization that embodies these sovereign characteristics. People from modern states prove their sovereignty not by practicing how they are in relationship with the Natural World but merely by belonging to this club.

The UN's self-constructed origin story aligns with Westernized ideals. One year after the UN Charter's ratification, Sumner Welles, a white male who served as a US State Department diplomat, delivered a lecture in October 1946, "The English Speaking Democracies." He spoke of the promise of a UN organization:

The nations of the world are engaged in an endeavor to fashion an international order through which humanity can obtain peace, and men and women everywhere can achieve the assurance of security, of liberty, and of ordered progress. As governments and organized groups search for the most effective means of achieving that ideal, they must necessarily seek to utilize in the structure to be erected those elements which have already proved themselves to be worthy and which can, because of actual human experience, be depended upon to support the stresses to which that new structure will be subjected, especially during the years of its initial growth (Welles, 1946, pp. 20-21).

Welles is, of course, referring to the Enlightenment (ordered progress) as the human experience and to democratic values (a social compact providing, among other things, liberty) as the requisite elements. But both are drawn from a Western perspective, which he expects the UN to universalize for all humanity. Welles knew full well that colonization and oppression persisted at the UN's birth. Examining the UN's fifty-one Founding Member States (FMS), self-determining peoples might rethink joining. For instance, of only three FMS from the African continent, one was South Africa, a modern state notorious for apartheid. European FMS had been busily colonizing both Africa's Indigenous Peoples, as an outcome of the 1884 Berlin Conference, and Southeast Asia. Moreover, twenty Latin American FMS, themselves on a path from being colonized to becoming modern states, nonetheless condoned genocide against Indigenous Peoples. Britain opposed Irish home rule, and the United States has continued its illegal occupation of the Kingdom of Hawaii since 1898. Yet, despite these and many other FMS violations of humanitarian, democratic values, Welles invoked the infamous writ large apologia, the white man's burden, in order to explain them away: "It was in part an outgrowth of that belief on the part of many of the English-speaking peoples that they must assume what was often termed 'the white man's burden'" (Welles, 1946, 5). In other words, whites get to determine for nonwhites how much civilization they can handle. Whites thereby control nonwhite self-rule, because only nonwhites who have internalized colonization are allowed into the club. With white oversight, nonwhites will monitor themselves and not regress to their non-Western, traditional lifeways.

Welles also knew full well that the United States, an FMS, had meat-bearing skeletons in its closet. Constitutionally sanctioned separate but equal doctrine normalized white American discrimination: Black Codes, Jim Crow, Japanese-American internment camps, and the unresolved

American question, Do Indigenous Peoples have the right of self-determination? To deflect UN critics by acknowledging humanity's shortcomings, adroitly, Welles mounted a critique of FMS and future non-FMS that subscribe to a master race ideology.

In a world in which even some of the oldest democracies have become permeated with the poison engendered by the doctrines of the "master race," and where the persecution and obliteration of millions of human beings because of their race or creed have, tragically enough, become almost common place, and where tolerance in certain of our English-speaking democracies is decried as either being impossible or even undesirable. . . . For no world order can ever be created upon the corrupt and disintegrating hatred which intolerance brings about. Discrimination against any race, or intolerance against any religion, can only lead to the eventual eruption of every one of those brutal forces which result in war, and in the suppression of all those individual freedoms in which we believe (Welles, 1946, pp. 22-23).

Welles's critique served at least two purposes. First, given the UN's FMS profile, his words reflected an aspirational goal. Because the master race ideology has yet to be disavowed, Welles's words veiled the UN's shaky foundation today. Second, FMS and other club members could shield themselves from forces, such as other ways of experiencing or knowing the world, that would challenge Western hegemony and its paradigm. The right to be a distinct people who base their sovereignty on non-Western principles, like recognizing the Natural World's personhood, could be targeted as subversive or dangerous to the West.

For instance, when confronted with a non-West sovereign that recognizes Natural World standing, as the Oceti Sakowin Oyate's Sacred Stone and Spirit Camps did, the United States, a FMS, condoned militarized police and a corporate mercenary campaign. But even the United States could not justify to the world the violence it arrayed against us (Matthews, 2016)[2]. Indeed, whenever "NoDAPL" broke through corporate-controlled media, US propaganda labeled Mni Wiconi action as an environmental protest at best and a threat to its nation (read: terrorism) at worst. To Welles I say "yes," a "world order" can exist on hatred at a high cost. The UN's inability to prevent its member states and others from resorting to violence shows a weak center, one that cannot hold.

Another indicator of US hatred toward Indigenous Peoples occurred sixty-two years after the UN's founding and nine years before the Sacred Stone Camp and Spirit Camp. In September 2007, the UN General Assembly approved the Declaration on the Rights of Indigenous Peoples. However, Canada, Australia, New Zealand, and the United States (CANZUS) rejected it and were the only FMS to do so.[3] While not a perfect document, the declaration contains provisions that "protect" non-Western ways of knowing. As our NoDAPL action shows, Indigenous nations reject development's underlying assumption: that the Natural World is an object and therefore rightless.

From Standing Bear to Standing Trees: The Unthinkable

The Sierra Club's 1972 federal lawsuit to protect a natural area from commercial development showed the West's resistance to recognizing other ways of understanding the world—not even Native in this case (Sierra Club, 1972). The Sierra Club lost its lawsuit because it did not have "standing." Standing is a legal—albeit human—concept. It means a person must have suffered a direct harm or wrong or be likely to suffer one from another person to have standing in a court. The Sierra Club's lawsuit failed because the club could not show a direct injury to it or any of its members from the development project.

However, before the court ruled to deny the Sierra Club standing, Christopher Stone penned a theory for the West: if nature is to be protected from development, it must be accorded legal standing. He contended that the Natural World, or at least its "natural objects," should be afforded personhood. The Western world view, of course, defines a person as a human being; but that definition has a checkered past. For example, the US Constitution counted Indigenous Peoples who had been taken from the African continent against their will and made slaves in a pre-1865 US society as three-fifths of a person. Peoples indigenous to North America, especially where the United States now illegally occupies, did not enjoy standing until 1879, when a federal judge issued a habeas corpus writ for Standing Bear, a Ponca leader, whom the US military had detained. The US argument against granting habeas corpus to Standing Bear centered on whether, like the Natural World, Indigenous Peoples qualify as persons. "In response to the writ, Mr. Lamberston, the United States Attorney, argued that by the very words of the habeas statute Congress had reserved the right to file the writ to 'those persons unlawfully detained.' Mr. Lamberston argued that Standing Bear was not a person because he was an Indian, therefore, he had no right to sue out the writ in a court of laws" (Nagle, 2012, p. 458).

Standing Bear v. Crook was asking nineteenth-century white society to consider the unthinkable: Indigenous Peoples are persons, that is, human beings who can suffer harms, injuries, or wrongs from others. What Stone proposed a century after *Standing Bear* about valuing the "rightless" for themselves rang loud in this trial. "Until the rightless thing receives its rights, we cannot see it as anything but a thing for the use of "us"—those who are holding rights at the time. . . There will be resistance to giving things "rights" until it can be seen and valued for itself; yet it is hard to see it and value it for itself until we can bring ourselves to give it "rights"— which is almost inevitably going to sound inconceivable to a large group of people" (Stone, 1972, pp. 8-9).

Before heralding the 1879 decision as white enlightenment, we must remember that little has progressed for Indigenous Peoples in the 138 years (and counting) since Standing Bear. Whites still defend "Indian" sport mascots, such as the Washington Redskins, Cleveland Indians, Kansas City Chiefs, or Edmonton Eskimos. Since most sports mascots draw from the animal kingdom,[4] Indian-as-mascot sends a clear message that whites— and those who agree with them—believe Indigenous Peoples are not human beings. The popular cowboys and Indians figurines (which anyone can purchase in the states today) send the same message: Native lives do not matter, because Natives are not fully human.

The US government institutionalized the message as well: the Interior Department is the site for Americans to negotiate their relationships with Indigenous Peoples. The Interior Department is responsible for the stewardship of trees, streams, lakes, "wildlife," minerals, land, and other natural wonders—the nonhuman presences to whom Stone proposed acknowledging personhood and standing. *Standing Bear* long since ruled that Indigenous Peoples are persons, hence rights holders. Why, then, is the State Department, where human relationships are primary, not the appropriate place for negotiating relations with Indigenous nations?

For Stone, Indians-as-mascots does not relegate Indigenous Peoples to rightless things, useful only for season tickets and a mass TV audience; it rather exposes whites' failure to recognize Indigenous Peoples as fully human. Whites' interest in maintaining Indigenous Peoples' rightlessness lies in maintaining their epic myth: our land was for them to freely take. The myth whitewashes their genocidal invasion. Indeed, as unconditional rights-holders, Indigenous Peoples demand to be made whole, that is, to have our pre-1492 self-determination actualized, however, much colonizers breach it. In addition to compensation, both land return and decolonization are the sine qua non for restoring Indigenous Peoples' wholeness.

However, the notions that Indigenous Peoples are human beings and that Mni Oyate are persons—unconditional rights-holders and not objects for whites to use—are anathemas to Western thinking: heresies. While Stone's theory for granting standing to nonhuman things faced incredulity from white society, the Water Protectors experienced far worse. Why? Because nineteenth-century, dehumanizing, and violent white doctrines—right of discovery, plenary power, and dependent domestic nations—still control Indigenous Peoples' fate. Mainstream society finds overturning these doctrines as unthinkable today as pre-1865white society found overturning slavery unthinkable or pre-1990 white South Africa found overturning apartheid unthinkable. The militarized police re/action at Spirit Camp and a US president's subsequent 2017 executive orders to authorize DAPL and other Western development projects proved the point: Americans still find it culturally unthinkable that Indigenous Peoples—not to mention the Natural World—are rights holders.

For the West, it is equally unthinkable to stand with the Natural World— in our case, with Mni Oyate—as a person, yet doing so affirms Mni Oyate as a holder of rights. Possessing rights means the Mni Oyate could, from a denial or violation of their rights, suffer a harm, injury, or wrong. To imagine this idea is to imagine that the Mni Oyate has legal standing. Stone argues that the West's anthropocentrism—or as he termed it, "the psychic and socio-psychic aspects"—throws self-interest in the way: why should humanity accept the Mni Oyate as a rights holder if it restricts how humans interact with water or the Natural World? In short, "What is in it for 'us'?" (Stone, 1972, p. 44).

With this reservation as to the peculiar task of the [human self-interest] Argument that follows, let me stress that the strongest case can be made from the perspective of human advantage for conferring rights on the environment. Scientists have been warning of the crises the earth and all humans on it face if we do not change our ways—radically— and these crises make the lost "recreational use" of rivers seem absolutely trivial. The earth's very atmosphere is threatened with frightening possibilities; absorption of sunlight, upon which the entire life cycle depends, may be diminished; the oceans may warm (increasing the "greenhouse effect" of the atmosphere), melting polar ice caps, and destroying our great coastal cities; the portion of the atmosphere that shields us from dangerous radiation may be destroyed (Stone, 1972, p. 45).

The culture shift facing the West boils down to how it defines a person, possessed of rights. Comparatively, the non-West continues stepping forward to making the shift: when human activity harms or injures specific natural objects or things, those harmed must be counted as persons with legal standing. For example:

In India, a court recognized Himalayan glaciers as legal persons, and the Ganges and Yamuna Rivers were given the same status as a human being. "This means legal guardians can now represent the waterways in court over any violation". In New Zealand, the Whanganui River is now recognized as a legal person. The Maori Nation fought nearly 150 years for the river to be recognized as an ancestor. In 2008, Ecuador built nature's rights into its constitution, ensuring that the country's entire ecosystem has a "right to exist, persist, maintain and regenerate" (John, 2017).

Responding to endangered, if not collapsing, ecosystems, these non-West efforts prove humanity can thread the needle of legal standing for the Natural World.

Otherizing the Natural World

The West, by contrast, defends only humans' rights. Almost two decades after Stone wrote, Joel Schwartz questioned the premise of nature as a rights holder: instead of considering the natural rights of persons who have legal standing because they have been harmed, he asked whether nature has the moral agency that rights require.[5] The ethical environmentalism that Stone and others advocated extends morality beyond the human circle to include nature. Schwartz believed this approach to be wrong: "Ethical environmentalism goes beyond Kantian morality by extending the applicability of the moral law, which no longer applies simply to relations among people, but also to relations between people and all natural objects. This expanded moral law asks people to regard as immoral the preference for human interests over those of animals, vegetables, and minerals. Anthropocentrism—the preference for our species over others—is held to be no less immoral than the selfish and unfair preference for oneself over other people" (Schwartz, 1989, p. 5).

Schwartz argued that human free agency and speech give us moral capacity, and consent is instrumental in having and exercising rights. The social contract or compact formalizes moral capacity. Human beings secure our rights, then, when we give our individual consent to form a sociopolitical body that protects our rights against the actions of others. Since the Natural World can neither "represent nor speak" for itself, Schwartz saw no moral imperative: "Stone simply assumes that nature, like the other entities that he mentions, can be represented; and even if it can be, he wrongly ignores the question of what entitles anyone to act as nature's representative. Nature cannot consent to being represented; it gives us no sense of how its "rights" are to be secured or who (since it cannot be what) is to secure them" (Schwartz, 1989, p. 6).

His words echo *Dred Scott v. Sandford* (Dred Scott, 1856). Dred Scott—though clearly a human being—had no legal standing. White law treated him as a nonperson object, denied the natural rights that his white, human owner enjoyed. It took a civil war among whites to break their hold on the idea—almost—that owning humans could be sustainable.[6]

For the West, recognizing the Natural World's personhood and rights raises the same issues that *Dred Scott* and *Standing Bear* raised: what view of reality grants a society to take from the Natural World without any thought of the harm that such taking causes? Schwartz's criticism goes to the core of how the West frames its experiences: no other reality matters, except that of humans.

Realizing Schwartz's argument cannot fully hold without qualification, Sandra Postel argues for a water ethic but within Schwartz's framework: human beings, not nature, are best suited to decide what is in nature's best interests. Her notion of human stewardship of the Natural World would involve a colonizing-like balancing act at best: "Instead of asking how we can further control and manipulate rivers, lakes, and streams to meet our ever-growing demands, we would ask instead how we can best satisfy human needs while accommodating the ecological requirements of freshwater ecosystems. ... Embedded in this water ethic is a fundamental question: Do rivers and the life within them have a right to water?" (Postel, 2008, A15). She tries to mitigate anthropocentrism's harms without changing the human-centrism that causes them. The Oceti Sakowin Oyate's response to her question is tentative: "Yes, but." We pause because she misses the nub: "Is water a rights holder?" As a people, we addressed this unequivocally in our action against DAPL: Mni Oyate does have rights, including the right to life, because Mni Oyate has personhood.

Western societies and their cultural heirs balk at our stand, though, because of how they see their place in the natural order. When the West engages in its global colonization projects, it Otherizes non-Westernized peoples. Otherizing uses socially constructed differences to claim that nonwhites are inferior, devoid of basic human rights. Colonizers use the construct to rationalize harming others for colonizers' benefit. With whites' hegemony at stake, the West institutionalized society-wide Otherizing that persisted well into our grandparents' and parents' generation: the United States' separate but equal doctrine, South Africa's apartheid, or Europe's anti-Semitism.

Colonization's Otherizing classifies non-Western peoples as nonpersons and therefore non–rights holders. In a colonizer framework, Otherizing implies that they, too, are objects or things no different from the things in nature (recall placing Indigenous Peoples under the US Interior Department's jurisdiction, not the State Department's). Stone's proposal challenged the West, as #NoDAPL did, to examine a core cultural belief. The West's inability to conceptualize the Natural World as little more than its playground has made it difficult for Western-based societies to grasp the Indigenous idea that the Natural World possesses personality. For example, a 1977 Haudenosaunee delegation to Geneva, Switzerland, told the world how Western development had altered the Natural World and that perhaps it was time for the West to rethink its core assumptions about the Natural World:

Western culture has been horribly exploitative and destructive of the Natural World. Over 140 species of birds and animals were utterly destroyed since the European arrival in the Americas, largely because they were unusable in the eyes of the invaders. The forests were leveled, the waters polluted, the Native people[s] subjected to genocide. The vast herds of herbivores were reduced to mere handfuls, the buffalo nearly became extinct. Western technology and the people who have employed it have been the most amazingly destructive forces in all of human history. Not even the Ice Ages counted as many victims. ...

The majority of the world does not find its roots in Western culture or traditions. The majority of the world finds its roots in the Natural World, and it is the Natural World, and the traditions of the Natural World, which must prevail if we are to develop truly free and egalitarian societies (Notes, 1981, pp. 76-77).

Serving development projects globally, colonizers from the West, including non-West societies who have adopted Western culture, have exploited every "object" or "thing" on Earth without a moral pang. They have targeted the Natural World for Otherizing regardless of the harms that follow. To persist in its course of mass harm and destruction, the West embraces a confounding paradox of anthropocentric thinking—that we can have it both ways. Mni Wiconi shows we cannot have Westernized water and drink it too. Coming to terms with the Natural World without having to change our behavior toward nature is folly of the deadliest kind. If we are to form sustainable relationships with/in the Natural World, then we have to live by the good relative mandate.

More Than Life

Postel observed that the much vaunted, globalized marketplace fails to value the most essential components of life. "Better pricing and more open markets will assign water a higher value in its economic functions, and breed healthy competition that weeds out wasteful and unproductive uses. But this [marketplace] will not solve the deeper problem. What is needed is a set of guidelines and principles that stops us from chipping away at natural systems until nothing is left of their life-sustaining functions, which the marketplace fails to value adequately, if at all" (Postel, 2008, A14).

Eons before Postel called for a global ethic to value the Natural World's life-sustaining functions rather than to use them for profits, the Oceti Sakowin Oyate lived—and, since our colonization, has struggled to live—by such an ethic. Phil Wambli Nunpa, Sicangu Lakota Treaty Council director, explained that "*Water is alive*: we call it mni wiconi, water is life" (Brave Heart Society, 2016).

The Oceti Sakowin Oyate acknowledge water as a distinct peoples—Mni Oyate (Water Nation). Of course, to claim that water is alive counters the West's construction of it as an inanimate object of nature. US law concerns water quality in order "to restore and maintain the chemical, physical, and biological integrity of the nation's waters" so that it is fit for human consumption" (US Code 33, 1251 et sq.). Water-usage rights are construed as either riparian or by prior appropriation. Water development projects serve flood control, irrigation, or hydroelectricity. All human-centered, these uses of water do not include a view of water as being alive, let alone conceptualizing it as a person possessed of legal standing. We, by contrast, recognize water as having personhood, independent of humans "giving" that standing or status, because of our *Otokahe Ka Gapi*: the story of First Beginnings or Creation. Through our story, fundamental values shape the Oceti Sakowin Oyate and our relationship with the Natural World.

> Inyan was in the beginning. Inyan began Creation by draining its blood to create. The first Creation was Maka, the Earth. After Maka, another need arose and Inyan drained its blood to address that need for Maka. As this [giving-of-self] process continued, Inyan grew weaker and weaker as its energy continued to flow into each Creation. ...
>
> Once Creation was complete, Inyan was dry and brittle and scattered all over the world. Today we use the Inyan oyate, the Stone People, in our inipi ceremony. . . . When the stones are brought in, we address them as tunkan oyate ("the oldest Creation Nation"). This [inipi ceremony] reminds us that the stones were in the beginning as Inyan.
>
> Through this story . . . we all come from one source, Inyan. We were all created out of Inyan's blood. To address all Creation as a relative, we use the phrase mitakuye oyas'in, "all my relatives" (White Hat, 1999, pp. 27-28).

Another term our people use to describe our shared Creation with the Natural World and universe is wakan. "Wa" is anything that is something, and that something possesses force-energy; "kan" describes a vein or channel from which something—such as force or energy—flows. Our Haudenosaunee relatives recognize a similar force-energy that they call *orenda*. For them, it is present in all natural presences, and these presences are fully capable of exerting it. This force-energy—wakan— and its power to flow are evident in the First Beginning when Inyan "opened all of his veins and his blood left him, and Inyan saw that his powers went from him in the blood and formed the edge of Maka" (Simms, 1987, 11). From Inyan's unselfish act of giving, he became the stones or rocks, that is, the dry and brittle material we encounter in nature. Significantly, how his blood transformed into the Mni Oyate, which then flowed around Maka, is how we come to know water as our relative.

In the Oceti Sakowin Oyate's world view, "ni," a root term found throughout D/L/Nakota vocabulary, expresses aliveness. Mni (water), Wiconi (life), Wicozanni (health), Woniya (life-breath), Inipi (steam purification), and Niya (an infant's first vital breath) together convey that water manifests life in our consciousness daily. For instance, George Sword shared his insight about the Inipi ceremony, illustrating how the ceremony not only reenacts the First Beginnings—when Inyan released a force-energy that flowed, enveloping Maka and imparting life to the Water Nation—but also binds the Mni Oyate and the Oceti Sakowin Oyate as intimate relatives.

> The white people call it a sweat lodge. The Lakotas do not understand it so. The Lakota think of it as a lodge to make the body strong and pure. They call it *initi*. . . .The *ni* of a Lakota is that which he breathes into his body and it goes all through it and keeps it alive. The spirit of the water is good for the *ni* and it will make it strong. Anything hot will make the spirit of water free and it goes upward. . . . An *initi* is made close so that it will hold the spirit of the water. Then one in it [the *initi*] can breathe it [the Mni Oyate] into the body. It [the Mni Oyate] will then make the *ni* strong, and they will cleanse all in the body. They wash it and it comes out on the skin like *te mini* [sic]. *Te mini* [sic] is sweat. It is the water on the body. A Lakota does not *inipi* to make the water on the body. He does it to wash the inside of the body.
>
> When a Lakota says *ni*, or *ini*, or *inipi*, or *initi*, he does not think about sweat. He thinks about making his *ni* strong so that it will purify him (Sword, 1991, p. 100).

Acknowledging that water has its own agency, which is personhood's attribute, causes Westernized folks to raise their eyebrows, as perhaps James Walker did when Sword explained this to him over a century ago. Yet recent water studies support Sword's claim. Dr. Masaru Emoto's water research challenges, if not defies, Western views about the Natural World as spiritless, inanimate objects. On one level, human bodies, on average, contain between 60 and 70 percent water by weight, but 99 percent of the molecules throughout the human body have water as a primary constituent. Therefore, normal body functions have a direct relationship with water. Dehydration reminds us of this relationship. But water has other qualities that suggest a much greater relational role than the West has imagined.

> "Studies of mindfulness, however, controversial, suggest that water responds qualitatively to intentionality. For example, an apple is about eighty percent water. Positive intentionality (good words and good thoughts) toward one half of an apple and negative intentionality (bad words and bad thoughts) toward the other half produce difference responses: the first half stays healthy, the second rots" (Valandra, 2016).

Since the human body, like an apple, is mostly water, both Sword and Emoto experienced water's responsiveness to intentions. Sword's observations that water has a spirit came from his ceremonial experiences and from our Creation story, while Emoto used technology to test what we already knew about the Mni Oyate. In both cases, intentionality provided the conduit in water's relationship with humanity. For example, one of Emoto's experiments subjected water in various containers to either harsh or euphonious music, offensive or nurturing language (either written or verbal), and negative or positive thoughts. He then froze the water to examine its crystalline structure. Was there any difference between water that had been intentionally treated well or badly? The results showed that water responds to intentions and reflects their qualities.

> All the classical music that we exposed the water to resulted in well-formed crystals with distinct characteristics. In contrast, the water exposed to violent heavy metal music resulted in fragmented and malformed crystals at best.

> But our experiment didn't stop there. We next thought about what would happen if we wrote words or phrases like "Thank you" and "Fool" on pieces of paper, and wrapped the paper around the bottles of water with the words facing in. . . .

Water exposed to "Thank you" formed beautiful hexagonal crystals, but water exposed to the word "Fool" produced crystals similar to the water exposed to heavy-metal music, malformed and fragmented (Emoto, 2004, xxiv-xxv).

Emoto's results are no surprise to the Oceti Sakowin Oyate and other Indigenous Peoples. From both Sword and Emoto, we can emotionally and intuitively begin to understand the relationship between these two peoples: humans and water. Sword described water's direct, salutary effect on us as extremely positive: water makes the Oceti Sakowin Oyate spiritually and physically strong from within, because love and gratitude are the two emotions we most often express in our lifeway through our prayers and songs. Hence, we integrate water's effect when we either breathe its vapor during an inipi ceremony or drink it. But the Mni Oyate also possesses another quality whose implication is significant: that water has memory.

In 1988, Jacques Benveniste (1935–2004), an immunologist from France, articulated a theory of water memory. Like Emoto's water studies and the Oceti Sakowin Oyate's understanding that water is alive, his theory remains controversial in the West. The theory states that water records, retains, and transmits information about its environment. Benveniste's theory explains that water molecules, through ionic attraction to one another, form closed structures —"structured memory." These structures have the capacity to record and store the electromagnetic signature, which means information about whatever object (or event) they come in contact with. Moreover, because the signature is electromagnetic, dilution does not affect the information's integrity. Hence, when we consume water in any form, water transmits to us the information it holds. With every drink or breath of water vapor, the Mni Oyate and human beings are bidirectionally communicating, however, unaware we may be that this two-way communication is going on.

The West continues to struggle metaphysically with what Indigenous Peoples have embraced since time immemorial, namely, the Natural World has personality and personhood, and nature's peoples are capable of having agency in relationships, such as mirroring emotions and even hashing things out. This deeper awareness of who the peoples of the Natural World are shows why Oceti Sakowin Oyate's action against DAPL is of a different order altogether. It comes from a far deeper understanding, from how we understand reality and the world we inhabit. As I explained in "We Are Blood Relatives":

That water is alive—and therefore possesses personality or personhood— defines our cultural response to DAPL. Our definition challenges the West's anthropocentrism, which accords person/peoplehood only to humans. Hence the Western way of life would both deny and defy water as having personhood. Yet the United States can arbitrarily recognize fictional entities like corporations as legal persons, while denying personhood to humans who become subject to the Thirteenth Amendment's slavery exception.

The Mni Oyate, then, is not unlike Indigenous peoples from Africa who, for 245 years in America, were racially constructed, socially viewed, institutionally handled, and economically exploited in the service of Western development. Similarly Western development frames water as a resource and as property: humans can own water and the [purported] right to harm water.

By contrast, our relationship with water is framed not as possessing rights over water but as protecting the rights of water. The Oceti Sakowin Oyate's original set of instructions requires us to be good relatives to the natural world. "What responsible conduct does water expect from us?" is a development question that we take seriously, however, alien the question is to ETP and those behind the DAPL project.

This culture-based understanding motivates the Oceti Sakowin Oyate to challenge the DAPL. How could it not? . . . During ceremonies such as the Sun Dance, individuals release blood from their bodies—as Inyan did at Creation—so that all life may continue on Earth. By weight, the human body is at least half water, making the Oceti Sakowin Oyate a blood relative of Mni Oyate. We are members of the Water Nation (Valandra, 2016).

Standing with Mni Oyate

When the Oceti Sakowin Oyate met modernity at Standing Rock, both the media (mainstream and social) and the West, responding as it did with state violence (militarized police and corporate security), missed the motivating core of our action. The media labeled our challenge to DAPL as environmental justice—a label that was not altogether incorrect. After all, ETP relocated DAPL just north of the Standing Rock Reservation after whites in Bismarck had objected to the pipeline as well, since it had originally been planned to cross the Mni Sose just north of their city. Clean water is an environmental concern, and oil spills contaminate water. But when the Oceti Sakowin Oyate argued precisely as the whites did about the pipeline's dangers, a militarized reaction rather than a route change was the response. Everyone knows why. Whenever a modern state pursues development, Native Peoples are expendable: our lives and losses are valued less than those of whites.

Unpacking the militarized police reaction to the Spirit Camp is complicated. One piece is the history of US violence against us. Ever since 1851, Americans have waged war against the Oceti Sakowin Oyate, requiring us to fight back in self-defense. We successfully outfought the US military in our homeland during the latter half of the nineteenth century, then again at Wounded Knee in spring 1973, and most recently in the latter part of 2016 when the Mni Wiconi call gained international traction. Another piece is our sovereignty and US treaties with us. When North Dakota and ETP managed through the DAPL decision-making process to marginalize a sovereign peoples' voice, the Oceti Sakowin Oyate asserted its sovereignty. The Americans' 1851 and 1868 Fort Laramie Treaties recognize our national boundaries, which DAPL crosses. But because of Americans' colonization and illegal occupation of our homelands, Americans willfully ignore their own treaties and trample on human, if not nature's, rights to life for the sake of Western development.

ETP's narrative about its decision-making process and the "public hearings" it held on DAPL invoked a process that did not include but marginalized our voices. Again, from "We Are Blood Relatives":

In its August 2016 progress report almost two years later, ETP stated the Dakota Access "has held 154 meetings with local elected officials and community organizations" and held "five public Open House meetings in North Dakota." To the uninitiated, Dakota Access comes across as a responsible corporate neighbor. What the progress report failed to mention, however, is that Indigenous peoples, such the Hunkpapa Titunwan (Standing Rock Sioux Tribe), were not included in these meetings. In their failure to understand that Indigenous peoples have a nation-to-nation relationship with the United States, Dakota Access and the US Army Corps of Engineers put Indigenous peoples in the position of having to react to the Corps' environment assessment after the fact.

While we must analyze how US colonization marginalizes our voices procedurally, the marginalization we experienced with DAPL is not the reason we take the stand we do. The Oceti Sakowin Oyate's and its allies' resistance to DAPL aims to protect water from harm (Valandra, 2016).

Controlling the national DAPL narrative was crucial for ETP, North Dakota, and other pro-DAPL stakeholders, for example, Wells Fargo, Royal Bank of Canada, and investors. Mainstream, corporate media colluded with state and corporate security forces to intentionally mislabel the NoDAPL action as violent and to imply that individuals at the camps were violent too, even implying that in our own homeland, we are terrorists. State and corporate forces used these insinuations to rationalize their militarized violence to shut down the Mni Wiconi action. But social media exposed the state and corporate lie. Social media recordings and other alternative media outlets at the Sacred Stone and Spirit Camps presented nonviolent, prayerful Water Protectors subjected to state/corporate intimidation and violence.

By complicating the corporate narrative, social media helped globalize NoDAPL and decentered state and corporate violence. Social media showed families, groups, and individuals arriving with supplies—food, water, medical, and other humanitarian items needed—from communities and organizations; they showed welcoming ceremonies when Indigenous nations arrived and posted their national flags to signify international solidarity; they showed the outpouring of goodwill, friendship, healing, and ceremonies at the camps; and they showed Indigenous leaders engaged in nonviolent decolonizing actions, even under the threat of state violence and incarceration. The images that went out over social media and other alternative news outlets left the West's institutions bewildered.

In contrast to the state's weaponized response, Indigenous Peoples conducted daily water ceremonies for the Mni Oyate throughout the camps. Our peoples showed a watching world that contrary to state and corporate press conferences and releases that framed the Mni Wiconi action as near terrorism, the Water Protectors engaged in prayerful relationships and peaceful activities among themselves and with the Mni Oyate. They modeled human beings who value water intrinsically, rather than as another resource to exploit. Water memory theory advises us to make peace with our relative, water—to recognize water as a legal person who is alive, who has agency and memory, and who holds legal standing.

Structural memory enables water to take an impression [or imprint] of anything that happens around it, and to connect all living systems together. And each one of us is a link in an endless chain of information transmission. But, in addition, each of us is also a source of information. Every one of our actions, a thought, an emotion, an uttered word separates from us and becomes part of the overall—and ergo—information environment.

Informational dirt is poisoning the water, accumulating layer by layer in its memory. If that process were to continue endlessly, the water could lose its mind. But [water] is endowed with a self-cleansing capacity. This occurs at the moment of phase transition, when it vaporizes and then condenses and falls as rain, or when it freezes and then melts. Shaking off the informational grime, water preserves its basic structure. That is the program for life (Perkul, 2009).

What is the Mni Oyate remembering of the humans at Standing Rock, and what stories will Water People transmit as they flow around the planet? What memories of Standing Rock will water eventually carry through our bodies too? We can only ponder. State militarized law enforcement and hired private, corporate security personnel stood against Water Protectors across an erected barrier, such as we see in fear-based societies uncertain of their legitimacy. They even used water to harm us with their water cannons, deployed in subfreezing temperatures. Threatening or engaging in violence is how humans—the presumed beings of moral agency—respond to nonviolent action in defense of the Natural World. The contrast between modernity's violence and traditional peacemaking could not have been more stark and clear to the Mni Oyate. Perhaps the Mni Oyate witnessed humans on the cusp of an existential epiphany.

Beyond political outcomes, the NoDAPL action created a watershed moment in human consciousness—maybe water appreciates puns too. Indigenous Peoples sheared away how the West thinks about the Natural World to reveal water as living peoples. Indeed, they are our relatives and we theirs. Our Original Instructions give us the responsibility to be in a good way with each other. That good way includes aiding the Water People when they are threatened. For the Oceti Sakowin Oyate and our allies, serving as water's protectors was and remains the right course, the one that matters most. Within our bodies, being a good relative is how we want the Mni Oyate to remember us.

Notes

[1] Valandra, E. (2019) Mni Wiconi: Water Is [More than] Life. In Estes N. and Dhillon J. (Eds.), *Standing with Standing Rock: Voices from the #NODAPL Movement*, pp. 71-89. Minneapolis, London: University of Minnesota Press.

[2] Alice Matthews, a Malaysian citizen, in September 2016 questioned President Obama about the Dakota Access Pipeline during a town hall youth conference in Laos.

[3] In May 2016 Canada fully supported, without qualification, the declaration. Australia, New Zealand, and the United States have officially supported the declaration with qualifications.

[4] The Indian-as-mascot apologists retort that other sports teams have human mascots (Minnesota Vikings, Tampa Bay Buccaneers, New England Patriots,

etc.). However, these mascots focus on an activity and are not racially objectifying. Anyone can be a pirate, patriot, saint, etc., but not everyone can be Native. See Living Justice Press's "Convention Style Sheet for Native Subject-Matters" http://www.livingjusticepress.org/

[5] The inalienable rights outlined in the whites' 1776 Declaration of Independence did not apply to the Indigenous peoples from the African continent, who became slaves, and North American continent, or to anyone who was not white, a male, propertied, Christian, and heterosexual.

[6] The much-heralded Thirteenth Amendment carves out an exception to involuntary servitude.

References

Brave Heart Society. (October 13, 2016). *Prayers and Actions Needed*. Available at: https://www.facebook.com/

Clean Water Act. (1972). US Code 33, 1251 et sq. Available at: https://www.epa.gov/laws-regulations/summary-clean-water-act#:~:text=%C2%A71251%20et%20seq.,quality%20standards%20for%20surface%20waters.

Dred Scott v. Sandford. (1856). 60 U.S. 393. Available at: https://supreme.justia.com/cases/federal/us/60/393/

Emoto, M. (2004). *The Hidden Messages in Water*. Hillsboro, OR.: Beyond Words Publishing.

John, T. (2017). The Legal Rights of Nature, *The Brief*. Available at: http://time.com/4728311/the-legal-rights-of-nature/

Kluger, R. (1976). *Simple Justice*. New York: Alfred A. Knopf.

Richard Kluger noted that "Americans more than most people tend to believe that progress must be upward linear, steadily and unbrokenly upward, or it is not progress at all". ((Kluger, 1976, 774).

Kolbert, E. (2014). *The Sixth Extinction: An Unnatural History*. New York: Picador.

Nagle, M. K. (2012). Standing Bear v. Crook: The Case for Equality under Waaxe's Law. *Creighton Law Review* (45)

Notes, A. (2005). *Basic Call to Consciousness*. Summertown, TN.: Native Voices.

Perkul, J. (2009) *Structured Water: Future of Medicine?* Available at: https://www.youtube.com/watch?v=dzptpMiVL1s

Postel, S. (2008). The Missing Piece: A Water Ethic. *The American Prospect*, 19(6), A15.

President Barack Obama on #NoDAPL and Dakota Access Pipeline. (2016). Available at: https://www.youtube.com/watch?v=gIMlc-iaxsk

Sierra Club v. Morton. (1972). 405 U.S. 727, Available at: https://supreme.justia.com/cases/federal/us/405/727/

Simms, S. (1987). *Otokahekagapi (First Beginnings): Sioux Creation Story*. Chamberlin, SD.: Tipi Press.

Stone, C. (1972). Should Trees Have Standing? Toward Legal Rights for
    Natural Objects.

Southern California Law Review. 45 (450).

Schwartz, J. (1989). The Rights of Nature and the Death of God. *Public Interest*
    (97).

Sword, G. (1991). Ni, Ini, and Initi. In DeMallie, R. J. and Jahner E. A., *Lakota
    Belief and Ritual*, p.100. Lincoln and London: University of Nebraska Press.

Valandra, E. (2016). We Are Blood Relatives: No to the DAPL. Hot
    Spots, *Fieldsights*, December 22. Available at: https://culanth.org/field
    sights/we-are-blood-relatives-no-to-the-dapl

Welles, S. (1946). *The Search for World peace, a Series of Lectures*. Winnipeg:
    Stovel company Ltd.

Chapter 6

Native Feminisms and Contemporary Indigenous Art: Emma Robbins and Angelica Trimble-Yanu Counter the Gendered Violence of Extractivism

Elizabeth S. Hawley

Introduction

In the customs of many Indigenous communities across the Americas, women are associated with and draw power from a deep connection with the Earth. Because of this sacred relationship, colonial invaders' violence against Indigenous lands via conquest, settlement, and development was— and is—inextricably tied to their brutal treatment of Native women. Today, legacies of this violence and brutality continue in the form of ongoing extractivist activities and their devastating environmental consequences as well as the high rates of violence against Native women that studies have shown to be particularly concentrated in areas of extractive industries. Native feminisms highlight the ways that Indigenous women are linked with the land–in Native customs as well as in continuing colonial abuses, albeit in very different ways[1].

This chapter presents the works of two contemporary Native artists whose practices draw from the intersections of their identities as Indigenous women, connections to their ancestral homelands, and avowed feminist positions. Emma Robbins's (Diné) mixed media works contend with histories of uranium mining on the Navajo Nation, the continuing effects of radon poisoning, and the roles of women in coping with resultant problems in Diné communities today.[2] Angelica Trimble-Yanu's (Oglala

Lakota Sioux Nation) prints, sculptures, performances, and videos evoke the Makȟóšiča (Badlands) and Ȟe Sápa (Black Hills) of her South Dakota homelands, symbolizing the significance of these landscapes for the artist as a Lakota woman and indirectly critiquing the pipeline construction that these lands and their peoples have been subjected to. Though the artists' practices differ in terms of the mediums used, Indigenous affiliations, and specific extractive industries addressed, these Native women are united in their commitment to countering the gendered violence of extractivism.

From the outset, I want to acknowledge my position as a non-Native feminist art historian. In this essay, I aim to privilege what Choctaw historian Devon Abbott Mihesuah terms the "reciprocal, practical dialogue" required for Indigenous voices and epistemologies to "become a part of feminist discourse," thereby aiming to practice the allied scholarship necessary for decolonizing feminist narratives and art historical canons (Mihesuah, 2003, p. 8). I am deeply grateful to the artists for conversations about the work discussed and for reviewing earlier drafts of this essay.

Native Feminisms

Robbins and Trimble-Yanu are members of a generation of early career Native artists who identify as feminists. The growing acceptance of the term *feminist* by some Native women is a relatively recent phenomenon; until about twenty years ago, many Native women rejected the term and the movement as unrelated and even antithetical to Indigenous worldviews and the contemporary concerns of Native peoples. The reasons for this disavowal were varied, but can be condensed into three primary, interrelated concerns: First, mainstream feminisms typically argue for gender equality, which is a concept that some Native peoples find to be at odds with Indigenous traditions of gender complementarity wherein men's and women's community roles differ but are nonhierarchically respected. Second, Native women in North America are dealing with the ongoing detrimental effects of colonization, and many women felt it expedient to privilege decolonization efforts over feminist ones. Finally, many Indigenous women viewed the field of feminism as whitewashed, observing that Euro-American feminist scholars assumed their own cultural frameworks to be universal. Indigenous feminists do not necessarily disagree with these assessments, but they have recontextualized them in ways that allow such critiques to coexist in productive tension with intersectional concepts of feminisms that address issues of gender and Indigeneity simultaneously (St. Denis, 2017, pp. 42–59; Hawley, 2022a, p. 30–32).

Extractivism and its devastating consequences—both historical and contemporary—loom large among these issues. Emphasizing Indigenous connections to ancestral lands and protecting the Earth from harm are central components of most Indigenous worldviews, and Native feminisms further underscore the key roles played by women in these aspects of Indigenous epistemologies. (It should also be noted that "women" is a category marked by an expansiveness extending beyond a binary system.) As English, Ktunaxa, and Cree-Scottish Métis scholar Joyce Green asserts, "Indigenous feminism draws on core elements of Indigenous culture—in particular, the nearly universal connection to land, to territory, through relationships framed as a sacred responsibility" (Green, 2017, p. 4). In many Native cultures, the land itself is considered to be a feminine force—Mother Earth, although different communities have varying and more specific names for this entity, and other female deities are linked to land in similar ways. Respect for Indigenous women and their leadership roles in their communities are innately connected to women's association with female spiritual beings, and, by extension, women's relationship with the Earth. Yaqui legal scholar Rebecca Tsosie observes, "The social and political power of Indian women was sanctioned by tribal religious traditions which often emphasized the role of female deities . . . Because the Earth's natural system depends on cyclical regeneration, the 'female' aspects of creation were particularly important" (Tsosie, 1988, p. 7). Or, as Laguna Pueblo writer Paula Gunn Allen succinctly states, "We are the land, and the land is mother to us all" (Allen, 1986, p. 209).

Emma Robbins: Addressing Legacies of Uranium Mining on the Navajo Nation

Diné historian Jennifer Nez Denetdale situates these connections between women and land within specifically Diné epistemologies, observing, "In Navajo life, mothers create and sustain life. Just as Changing Woman sustained her children, so do all Navajo women today. All things that sustain life, including the Earth, agricultural fields, corn, sheep, and even men who are nurturing, are called mother. Furthermore, motherhood is the role from which Navajo women speak with authority" (Denetdale, 2007, p. 137). This wide-ranging vision of motherhood and the Diné creation stories from which it originates continue to guide Diné approaches to the lands of the Navajo Nation today.

Denetdale's reference to Changing Woman, or Asdzą́ą́ Nádleehé, is significant; she is a prominent figure among the Diné Holy People. Changing Woman is in many ways a personification of the Earth herself, responsible for the changing seasons, the birth of the original clans from whom all Diné are descended, and Diné women's transition through puberty. Her role gives a sense of the importance of women's positions and responsibilities in Diné culture (one in which the community duties of women and men are equally respected, and the parameters separating them are permeable) as well as the interconnectedness of all things in Diné worldviews, wherein the health and well-being of humans is inextricable from the state of all other entities. Changing Woman, moreover, through a cosmic union with the Sun, produced offspring known as the Hero Twins—Naayéé' Neizghání (Monster Slayer) and Tóbájíshchíní (Born for Water)—who rid Dinétah (Diné homelands) of monsters in an ancient world prior to this one. As art historian Janet Catherine Berlo notes, such stories of the ancients are indissociably linked to "new ecological 'monsters' associated with industrial resource extraction, including devastating pollution caused by nearly four decades of uranium mining" (Berlo, 2009, p. 245).

The unleashing of this new monster occurred during World War II, and the havoc it wreaked continues to devastate Diné and Dinétah—bodies of the people and bodies of land that cannot be separated one from the other. Uranium mining was carried out on the Navajo Nation—a reservation located on portions of present-day Arizona, New Mexico, and Utah—from 1942 to 1986, initially under the auspices of the Manhattan Project and subsequently for the Atomic Energy Commission (AEC) established by the postwar congressional passage of the 1946 Energy Act. Due to uranium contamination, homes, crop soil, livestock, and water sources continue to have dangerously elevated levels of radiation, causing an array of health problems that can prove deadly.

Figure 6.1. *Emma Robbins, What You Should Know About Radon, 2020. Brochures, thread, 24.5 x 21 in. Image courtesy of the artist.*

There are significant examples of Diné men and non-Native artists who have visually grappled with these monsters—including Diné photographer Will Wilson, whose work is included in this publication—but Diné women have been at the forefront of aesthetic and activist responses (Hawley, 2022b, pp. 66–71). Emma Robbins is among these women, and she directly addresses concerns regarding uranium mining in *What You Should Know About Radon*, a wall hanging that features leaflets of the same title sewn together in the creation of a composition that evokes historical Diné weaving practices even as it operates in the contemporary art arena of found object assemblage (Fig. 6.1). The multimedia work emphasizes the dangers that elevated radon levels caused by uranium mining pose, threading together pages from pamphlets issued by the Navajo Nation Environmental Protection Agency (NNEPA) that define radon as a radioactive gas that "comes from the ground and is found in minerals and soils that have high uranium decay, which produces Radium, then Radon;" describe where it might be found: "It travels upward through the soil and into your homes and buildings through cracks in the floor and walls, openings around pipes, and the water supply;" and outline associated health risks, emphasizing in particular how the inhalation of radioactive particles can lead to lung cancer (Navajo Nation Environmental Protection Agency, n.d.,b).

Robbins gathered these pamphlets from the Naatsis'aan (Navajo Mountain) Chapter House.[3] The brochures are undated, but the NNEPA was founded as the Navajo Environmental Protection Commission in 1972 (taking its current name and organization in 1995), by which point leaders of the Navajo Nation had grown increasingly concerned about the health hazards associated with uranium mining. Mining had expanded on the reservation during the Cold War with little care given to the mostly Indigenous workers and their families. The US government put a premium on locating domestic uranium deposits because enriched uranium oxide powder can be used to fuel nuclear power plants, and, with additional processing, can also be employed as the fissile core of nuclear bombs, making this material a key resource during World War II and the nuclear arms race with the Soviet Union that followed. When significant amounts of the ore were found on Diné territories, federal officials pressured the Navajo Nation Council into leasing the land to mining corporations.

Struggling with widespread poverty, misled into believing the mining operations would stimulate economic development, and with no information given about the known health hazards of such extractions, council members agreed. (Johnston, Dawson, and Madsen, 2007, pp. 97–101).

Over the next four decades, more than 3,000 Diné were employed in some 2,500 mines in and around the Navajo Nation; damaging Diné and Dinétah alike with health and environmental problems (Voyles, 2015, p. xiii). Today, the NNEPA characterizes the harms done to the Navajo Nation in terms that echo the connections Native feminists have made between women and the Earth; in a quote pulled from a resolution of the Navajo Nation Council (CAP-47-95), the landing page of the organization's website states, "In the Navajo Way, the Earth is our Mother, the mountains part of her sacred body, the water courses her veins and arteries. When the earth is injured, the resultant instability, imbalance and disharmony bring illness to life on Earth including mankind" (Navajo Nation Environmental Protection Agency, n.d.,a).

Recent scholars have also underscored the experience of women who may not have worked in the mines themselves but were nonetheless subjected to the harmful effects of the practice, which have included radiation-related cancers and other diseases, respiratory ailments, and reproductive health issues. Women's studies scholar Traci Brynne Voyles reminds readers that while the workers themselves were mostly male, their female relatives were endangered by the contaminants in the course of completing community tasks associated with their gender—"women were exposed to radioactivity when men came home from work covered in radioactive mud and dust; when they laundered workers' clothes; when radioactive dust settled onto the swept dirt floors of their hogans, where children played; and when they slaughtered, prepared, and ate contaminated livestock" (Voyles, 2015, p. 135). Miners' families also "reported that they used water from the area for drinking, bathing, washing, and household uses" (Johnston, Dawson, and Madsen, 2007, p. 106).

Lack of access to clean water continues to be a major problem on the Navajo Nation, and it is a concern that has been at the forefront of Robbins's work for several years. Robbins is the definition of an artist-activist; she has put her art practice partially on hold after taking a position in 2016 with the human rights nonprofit DigDeep, where she currently serves as the Navajo Water Project Executive Director. As stated on the Navajo Water Project website, one in three Diné community members live in a household without running water, an environmental injustice directly related to broken treaty promises made by the US government over 150 years ago. The Navajo agreed to relinquish vast swathes of their land in return for US assistance with housing, infrastructure, and healthcare on the reservation, but the US reneged on their end of the agreement (Navajo Water Project, n.d.). The Navajo Water Project is a community-managed organization aimed at fulfilling those unmet needs even as they highlight the federal government's failed responsibility to do so.

Robbins draws direct connections between her work with DigDeep, uranium contamination on the Navajo Nation, and loss within her own family and the greater Diné community. In an artist talk for one of my undergraduate classes, she shared, "The way that I actually became interested in working in water and with the environment was because I lost my grandmother when I was fourteen to stomach cancer [due] to radon contamination in her water source. And my story is not unique; that's something that a lot of us have experienced . . . loss or illness." She notes that abandoned mines on Dinétah—left with little to no efforts at cleanup after the mid-1980s—"are not labeled well, and a lot of people who don't have running water might be getting water from these contaminated sources" (Robbins, 2022).

Robbins, moreover, points to the burden that the lack of a clean water source places on Diné women, who have historically drawn power and leadership responsibilities from their position as caretakers of the home. In an artist talk for the 2021 apexart exhibition *Native Feminisms*, she emphasized how Diné women "have been the leaders of our communities, and we are the caregivers of our communities," clarifying that "it's not the idea of the 1960s, stay in the kitchen, take care of your family," but rather characterizing the home as a potent platform from which to provide for kin (Robbins, 2021). When the home is compromised, which is certainly the case when clean water is unavailable, Diné women struggle to fulfill caretaking responsibilities rendered unfeasible by these environmental hazards. Voyles links such situations to environmental sexism, which she argues "occurs when women's roles as caretakers compound the burden of environmental problems in their lives." She elaborates: "It is women who take up the labor of care when family members become sick; it is women who often assume doubled financial responsibilities when their husbands or partners die" (Voyles, 2015, p. 142). For Indigenous feminists, the problem is not that women are expected to serve as caretakers—this is a revered and respected role within most Native communities—but that they are having to do so today within untenable situations negatively impacted by legacies of colonialist policies that include extractivist industries.

For Diné women, another customary role involves weaving, which Robbins evokes with the threads she uses to underline portions of text and stitch the pages of the brochures together in *What You Should Know About Radon*. Weaving conventions are connected to the very origins of Diné culture; Na'ashjé'íí Asdzáá (Spider Woman) wove the web of the universe, connecting all entities via her cosmic threads and giving women the tools needed to hold Diné society together. Anthropologist Kathy M'Closkey connects this all-encompassing worldview of weaving to the environment, describing weaving as "part of an ecological nexus" and citing a Navajo philosopher who states, "What the women weave is part of the environment . . . If you take something from the environment, you must give something back. Navajo weaving is all about relationships . . . we are like children in our relation to Mother Earth" (M'Closkey, 1998, pp. 121–22). The reciprocal relationship between Diné and Dinétah, however, is ruptured by extractivist activities propelled by nationalism and capitalist concerns driven by profit rather than community well-being. Robbins's stitches might thereby be read as sutures threaded through the pamphlets in an attempt to heal the ailments highlighted therein.

The crimson color of these threads is also highly evocative, signifying a visual connection with the red earth of Dinétah as well as the violence connoted by this hue—links that are interrelated, as Native feminists have argued, given women's associations with the Earth. Muscogee [Creek] Nation legal scholar Sarah Deer, for example, asserts that Indigenous Peoples experience "intrusion on their lands and culture by an exterior, hostile outsider," whereas "rape victims experience the same dynamic, but it is played out on their bodies and souls rather than the land." She concludes that due to the long history of sexual assault against Native women, "rape can be employed as a metaphor for the entire concept of colonialism" (Deer, 2015, pp. xiv, xvii). As I have written elsewhere, "While Indigenous worldviews that connected women with the earth operated from an attitude of respect and veneration, Euro-American epistemological connections between Native women and the land they inhabited manifested in their sexual subjugation" (Hawley, 2022a, p. 39).

Settler communities justified this subjugation in part by characterizing Native women as sexually impure and therefore available. Early invaders found that some Indigenous communities practiced polygamy, and some allowed for sexual relations prior to unions that Euro-Americans might term *marriage*. Instead of considering that this behavior took place within entirely different communities and worldviews than their own, male settlers imposed their own, typically Christian biases regarding sexual sin on Native women. As ethnic studies scholar Andrea Smith contends, "in patriarchal thinking, only a body that is 'pure' can be violated. The rape of bodies that are considered inherently impure or dirty simply does not count," and she characterizes the settler fantasy of Native women's immanent sexual availability part of the "sexual colonization of Native peoples" (Smith, 2003, p. 73).

Figure 6.2. *Emma Robbins, Breast Implants, 2013. Blue Bird Flour bags, bleached flour, beads, thread, dimensions variable. Image courtesy of the artist.*

This harmful fantasy continues to imbue settler American visual imagery of Indigenous women. As Deer observes, "Today, the eroticized image of Native women is so commonplace in our society that it is unremarkable—the image of a hypersexual Indian woman continues to be pervasive in American culture," giving examples that include "Poca-hottie" Halloween costumes and fetish porn images of the "rez girl" (Deer, 2015, p. 62). In some of her photographic work, Robbins has staged herself in costumery associated with these stereotypes, reappropriating the aesthetic of the hypersexual Native woman to underscore its fallacy. But perhaps her most astute work in this area is *Breast Implants,* a piece that features scraps taken from bags of Blue Bird Flour—a product ubiquitous on the Navajo reservation where this flour is primarily sold—that feature the blue birds of the company name perched on stacks of the wheat from which the flour is ground (Fig. 6.2). Robbins has sewn these pieces into two convex domes punctuated by blue beadwork at the centers—creating breasts and nipples whose artificiality is underscored by the title of the work as well as the lush greenery and clear blue skies of the landscape that replace flesh tones.

Breast Implants deftly highlights Native women's homemaking tasks and nurturing links to the land, even as it underscores imposed settler attitudes regarding sexual and geographic conquest. The piece is typically Diné in its tongue-in-cheek humor, but it nonetheless alludes to the grim consequences that connections between bodies of land and bodies of women have had for Native women. As Voyles notes, the influx of Native and non-Native workers to mining boomtowns led to a marked increase in cases of rape and domestic violence (Voyles, 2015, p. 147–48). Robbins's work visually dovetails with Voyles's assertion that Indigenous women "situate the uranium industry in a national context of sexually violent colonial practices, where the frontier, like the concept of 'Nature' in general, is constructed as feminine, and colonial ventures into it are 'penetrations' that can be understood as deeply sexualized acts of violence against the natural environment and indigenous people alike" (Voyles, 2015, p. 149). It also distinguishes her as a key advocate in the ways Voyles argues Native women are "framing uranium mining as one facet of reproductive injustice," and insisting that environmental justice be considered "not merely as an issue of the distribution of environmental harm, but as evidence of a much larger, systemic problem . . . of the deeply intersectional nature of race, gender, and reproduction in colonization for women" (Voyles, 2015, p. 144). Voyles notes that it is from this intersectional condition that Indigenous women have taken up leadership positions in decolonizing activist movements. Robbins is in good company.

Angelica Trimble-Yanu: Protecting Lakota Relatives from Pipeline Violence

In recent years, the extractivist "penetration" that has gained the most mainstream attention is the Dakota Access Pipeline (DAPL), a transport system for moving crude oil from the Bakken oil fields of North Dakota through South Dakota and into Illinois. Representatives from Energy Transfer Partners (ETP), the Texas-based company overseeing the construction of the pipeline, met with Standing Rock Sioux Tribe council members in 2014 to seek their approval for the project. The council members rejected the plan, expressing particular concern that potential leaks in the DAPL would contaminate the community's water supply and that the pipeline construction process would desecrate sites sacred to the Lakota and Dakota peoples who comprise the Sioux (Gilio-Whitaker, 2019, p. 1–3).

ETP ignored these concerns and broke ground for the DAPL construction in 2016, unleashing Zuzeca Sapa, the Black Snake that a Lakota prophesy warned would bring destruction to the land. The massive resistance movement that would eventually become synonymous with Standing Rock began on a small scale; a handful of women from the Standing Rock Sioux Tribe set up a camp they dubbed *Camp of the Sacred Stone* (or *Sacred Stone Camp*) from which they intended to monitor construction practices, continue to register their tribe's dissent, and, they hoped, convince ETP to reroute the lines away from the community. Instead, ETP bullishly moved forward, mounting a campaign of media misinformation regarding the tribe's concerns while simultaneously bringing in armed security personnel to deter the peaceful protectors. LaDonna Bravebull Allard—a tribal historian of the Standing Rock Sioux who had been one of the Camp of the Sacred Stone's original organizers—put out an open call inviting supporters to camp out on her land in order to physically take a stand against ETP's pipeline. As news of the situation grew, thousands of Native and non-Native allies from across the nation and world poured in to #Standwithstandingrock, per one of the hashtags through which the situation gained attention on social media. Others sent financial support, donated supplies, and shared messages of solidarity. As Peoria historian Elizabeth Ellis avouches, the conflict kindled "a fire burning in the heart of Indigenous America, and it has ignited a generation of activists and forged broad solidarity unlike any we have ever seen" (Ellis, 2019, p. 172).

And it all began with Native women, who remained central to the Standing Rock movement. As Sisseton-Wahpeton Oyate Native studies scholar Kim TallBear observes in an essay connecting this movement to earlier women-led social and environmental initiatives, "badass indigenous women caretake relations" (Tallbear, 2019, p. 13). *Relations* is a particularly significant term, and one for which many Native languages have a roughly translatable word. In the Lakota language, the phrase *Mitákuye Oyás'iŋ* (all are related or all my relations) refers to a concept central to Lakota worldviews: that all entities are connected in a universal oneness, be they humans, animals, plants, lands, waterways, and so on, and harmony is achieved by being a good relative to all the beings with which one is surrounded. Women, moreover, are the caretakers of these multiple beings; ensuring the welfare of relatives is a customary and deeply respected female role in Lakota culture, stemming from their positions as mothers and keepers of the thípis. That women took up leadership roles at Standing Rock comes as no surprise, since, as anthropologist Marla Powers comments, "in Lakota culture maternal and managerial roles are not regarded as antithetical" (Powers, 1988, p. 3). Caretaking comes in many forms.

Figure 6.3. *Angelica Trimble-Yanu, Iyeska, 2018. Oil monotype, 30 x 22 in. Image courtesy of the artist.*

Angelica Trimble-Yanu was not physically present at Standing Rock during the DAPL controversy. Born in Oakland, California to a Lakota father and non-Native mother, she spent her early years between Oakland and the Pine Ridge Indian Reservation in South Dakota. Complicated family circumstances eventually led to her being adopted as a toddler by her maternal grandmother who raised Trimble-Yanu in Oakland. She returned to her ancestral homelands for the first time as an adult, in 2018, during the making of *Iyeska*, a multimedia project that variously incorporates monotype prints, site-specific sculptures, performances, installations, and documentary photos and video (Fig. 6.3).[4] It was also on this visit that she reconnected with her relations—all of them, including both her father's family members and the Makȟóšiča (Badlands) and Ȟe Sápa (Black Hills) sacred to Lakota communities. She characterizes this trip and the artwork that spurred it as a "return to my ancestors and our land" (Trimble-Yanu, 2022).

The title of the piece speaks to the artist's liminal status as being both of and apart from her people; she is a biracial woman, spent early years on the reservation but was primarily raised elsewhere, is deeply committed to her ancestral homelands from afar, and uses her art to underscore the importance of Indigenous ontologies for audiences that include many non-Native viewers. *Iyeska* roughly translates to *interpreter* and it is a term that once referred to a well-spoken person. It took on a new meaning when Lakota communities began interacting with non-Native groups who needed translators who could understand and speak both the Lakota and English languages. The community members who filled this role—one that dovetails with historian Margaret Connell Szasz's characterization of *culture brokers*—were labelled *Iyeska* (Szasz, 1994, p. 19–20). Trimble-Yanu's own uncle, Charles Trimble, served in this role as Executive Director of the National Congress of American Indians in the 1970s, and it was his book *Iyeska* that first introduced Trimble-Yanu to the term (Trimble, 2012). But the word also held some derogatory connotations of "half-breeds" and "traitors" who were not "real Indians" (Burnette, 2018, p. 45). In titling her work this way, Trimble-Yanu reappropriates the word, insisting on what Ojibwe and Dakota scholar Scott Richards Lyons has called *rhetorical sovereignty*, which he defines as the "inherent right and ability of peoples to determine their own communicative needs and desires in the pursuit of self-determination," further asserting that "rhetorical sovereignty requires above all the presence of an Indian voice, speaking or writing in an ongoing context of colonialization. Ideally, that voice would often employ a Native language" (Lyons, 2000, p. 462).

Trimble-Yanu began creating the black and white monotypes of *Iyeska* in 2016, many months before she landed on the title for the project and before she planned her journey home. She describes her initial drive to make these prints as an almost obsessive working through of her memories of her homelands. The striking monochromatic compositions do not represent specific locations, but rather abstractly recall the striated geographic formations of the hills that swell from these lands. Black and white are colors sacred to the Lakota, but the inky black sections of the works also might be seen as an evocation of the oil being pumped through pipelines that zigzag through Native lands across the US—an evocation made all the stronger by the fact that the Standing Rock movement began in the same year. That Trimble-Yanu felt compelled to create these pieces during the same months that fellow Sioux community members and allies were seeking to protect Ȟe Sápa from yet another incursion of oil circles back to Mitákuye Oyás'iŋ. All are related, and the artist's compulsion to create print after print of imagery that speaks to her absence from Lakota lands even as it insists on the lands' deeply felt presence in her mind and memory at a time when these territories were under attack manifests the characteristic dualities and cyclical spatiotemporalities of Lakota epistemologies.

In discussing *Iyeska*, Trimble-Yanu observes, "Typically, the western gaze is extractive, and they often only ask what will this land give me versus what can I give this land" (Trimble-Yanu, 2022). This assessment might describe nineteenth-century non-Native painters who profited from depicting sweeping western vistas and early conservationists who seized ancestral Indigenous lands for national parks, and it is certainly an accurate characterization of the voracious oil and other extractive industries of today. This land has been mined in many ways by forces that have seen it as a resource rather than a relative. Trimble-Yanu thus gave a lot of thought to how to return to her homelands after having been away for so many years, asking how she might "return in a good way, an ethical way," acknowledging that "even though I came from this land, I didn't know this land. I hadn't given an offer to these spaces yet" (Trimble-Yanu, 2022).

Figure 6.4. *Angelica Trimble-Yanu, Iyeska, 2018. Documentary photograph. Image courtesy of the artist.*

Her offering came in the form of her art, which is as much ceremony as it is practice, linking her to relatives in a multitude of ways and underscoring her identity and responsibilities as a Lakota woman. She took her monotypes into the hills, shaping them into paper sculptures that she nestled into the landscape and photographed (Fig. 6.4). She emphasizes the collaborative and maternal aspect of the process, crediting the mother, grandmother, and auntie figures who traveled with her, relayed stories to her, and helped her document the making—or birthing—of *Iyeska*. Photographs show Trimble-Yanu at work in the landscape, which the artist acknowledges as another contributor: "I sculpted with the landscape—meaning I let the spaces inform the structure of the final piece" and she emphasizes the "embodied knowledge" engendered by the project (Trimble-Yanu, 2022). Body of work, body of knowledge, body of land, body of woman—*Iyeska* embodies a multitude of relations.

Trimble-Yanu underscores "the importance of me returning there and making that work there that connects my body as a woman to that land," further noting that, "I really learned the importance of being a woman in these spaces and how we reconnect to these spaces and bring healing back to our people. And there's just this medicine that happens when women collaborate with the landscape, especially our Lakota women" (Trimble-Yanu, 2022, personal communication, 30 December). Her deepening understanding of her identity as a Lakota woman and the significance of the culture bearing with which this role is associated was encouraged by women in her Lakota family and community, who spoke with her at length, sharing their stories. As journalist Sarah Penman reports, "In traditional Lakota and Dakota society, grandmothers were respected for their knowledge, wisdom, and power as life-givers, healers, dreamers, harvesters, and teachers . . . they counseled girls on their moral, social, and spiritual responsibilities," and even today "it is the grandmothers who shoulder the task of maintaining traditions and retelling the old stories — not simply as memories from the past, but as living elements in the contemporary American Indian experience" (Penman, 2000, p. 3–4).

The powerful community role of women draws in part from the legacy of Ptesáŋwiŋ, or White Buffalo Calf Woman, a significant spiritual figure for the Lakota, and one who Trimble-Yanu frequently references in regard to her work. Many versions of her story have circulated, but the central narrative is an account of a long-ago time of famine, during which two young men came upon a beautiful woman while hunting. One of the men was filled with lust and declared that he would take the woman back to his thípi. He tried to embrace the woman, after which a cloud enveloped him, and he disappeared. The woman then instructed the other man to return to his community and have them prepare a feast for her, which he did. When the woman arrived, she brought with her the Buffalo Calf Pipe. With the pipe, she explained that Lakota would be connected to the ground beneath their feet and the skies to which the smoke reached, becoming one with all entities in the universe in a state of Mitákuye Oyás'iŋ (Walker, 1991, pp. 109–12; Hämäläinen, 2019, pp. 164–66). She also outlined gender roles; as one Lakota elder recounts, "The White Buffalo Woman then addressed the women, telling them that it was the work of their hands and the fruit of their wombs which kept the tribe alive. 'You are from mother earth,' she told them. 'The task which has been given you is as great as the one given to the warrior and hunter.' And therefore, the sacred pipe is also something which binds men and women in a circle of love" (Lame Dear and Erdoes,

1972, pp. 253–54). The woman transformed into a white buffalo calf upon leaving the camp, and the plains were suddenly swarming with the buffalo herds that sustain Lakota peoples.

This origin narrative is significant in myriad ways, but of particular note for Native feminisms is the prohibition of sexual assault against women, as symbolized by the White Buffalo Calf Woman punishing the lewd man in the narrative. Historical accounts of domestic abuse and assault are rare in the preinvasion era of most Indigenous Peoples, and Lakota social systems had protocols in place that prevented and protected women from such behavior (Reyer, 1991, p. 71). But with colonization came abusive attitudes toward Native women, as outlined in the previous section of this essay. Today, the "man camps" that spring up around areas of extractive industries lead to an increase in cases of violence against Indigenous women. A primary concern of the DAPL protectors was the influx of non-Native workers that the pipeline construction would bring the area and the known increases in gender-based violence that would result—violence already suffered by the Indigenous women who live near the Bakken oil fields from which the DAPL would transport the crude product. As Cherokee playwright and lawyer Mary Kathryn Nagle and noted feminist Gloria Steinem observed in an op-ed: "In North Dakota, the man camps created during the Bakken oil boom drastically increased the levels of violent crime perpetrated against women and girls—and particularly native women and girls. Studies conducted during the peak of the oil boom—from 2010 to 2013—showed that the number of reported domestic violence incidents and sexual assaults increased by hundreds, flooding and overwhelming service providers." They underscored how "the Mandan, Hidatsa, and Arikara Nation . . . became ground zero for the increase in violent crimes that accompanied the boom" which included "domestic violence, sexual assault, and sex trafficking" (Nagle and Steinem, 2016).

Figure 6.5. *Angelica Trimble-Yanu, Iyeska, 2018. Glass, clear fishing line, oil monotypes on Mohawk paper, hanging sculpture 1: approx. 37.5 x 17.5 in., hanging sculpture 2: approx. 31 x 23 in., hung at variable heights above 4 ft. diam. mirror. Image courtesy of the artist. Photo: Mario Gallucci.*

Trimble-Yanu's experience in the creation of *Iyeska* and the aesthetic choices she made in its installation speak to this threat of violence. She recalls "having experiences that felt unsafe" while driving through South Dakota, and she connects the red thread from which her paper sculptures hang when *Iyeska* is presented in gallery settings to both women's customary art-making practices as well as MMIW (Missing and Murdered Indigenous Women, sometimes extended to MMIWG2S, or Missing and Murdered Indigenous Women, Girls, and Two-Spirits), a grassroots movement aimed at highlighting the horrifyingly high rates of violence suffered by Native women (Fig. 6.5). The artist notes that historically, women were the community members responsible for the sewing and beadwork involved in the creation of regalia, and she stresses the "material choice" she made in using string of "the color red, which is seen in a lot of our regalia." The rounded mirror she places beneath the sculptures also underscores womanhood; she has variously compared its shape to a womb, the sun, and the circular imprint on the Earth left after women's dances, with the reflective surface symbolizing Lakota connections between Earth and Sky prescribed by White Buffalo Calf Woman. But Trimble-Yanu observes that "more recently, in the last couple of years, I've seen red being used as a symbol for voicing the issue around Missing and Murdered Indigenous Women" and she has come to view the hue in this way as well (Trimble-Yanu, 2022, personal communication, 30 December). She further explains that "the color red is also a way of calling back the spirits, it's a form of grounding and connecting. We believe it's the color that our ancestors are naturally drawn to. It's a form of calling back . . . the souls of those harmed to allow them to cross over to the spirit world" (Trimble-Yanu, 2023, personal communication, 3 January).

Robbins has also connected her work to the MMIW movement, remarking that "these are our sisters . . . and our mothers and our grandmothers and our aunts . . . when they go missing from our community, we're losing a big part of who we are" (Robbins, 2021). The red sutures of *What You Should Know About Radon* function in much the same way as Trimble-Yanu's string, albeit drawing from a culturally distinct Indigenous community. While accounting for such distinctions is an important aspect of decolonized scholarship, so too is acknowledging instances where communities have come together in solidarity to fight against colonial forces; the interrelated MMIW and Standing Rock movements are just two recent examples.

Enduring Relations

Though the artists addressed in this chapter come from two different Native communities, the machinations of and resistance to extractivist industries link Diné and Lakota histories, feminisms, and embodied experiences. Voyles reminds readers that uranium mining also occurred on the Pine Ridge reservation, and that Indigenous women activists in the 1970s linked this to reproductive health risks, pointing to "the kinds of birth defects, reproductive anomalies, and spontaneous abortions (miscarriages)" that were also "reported among Navajo and Pueblo women in the Southwest" (Voyles, 2015, p. 141). At Standing Rock, over 360 Indigenous nations joined the resistance against the DAPL, and Diné protectors—which included Robbins—are among those emphasized in accounts of the movement. Ellis, for example, writes, "Navajos, and so many other Native Americans perceive the threat of DAPL not just as a danger for the Standing Rock Sioux, but as a critical threat to Indigenous sovereignty and resources." She then links these concerns to the treatment of women, asserting, "Moreover the exploitation of Indigenous resources frequently enables violence against indigenous women, and so the threat of colonial resource extraction resonates with communities who have suffered gender-based violence" (Ellis, 2019, p. 175).

In one of the most evocative accounts, Diné social scientist Teresa Montoya describes a fitful dream she had while journeying to Standing Rock to join the protectors. She recalls having visions of a desert landscape she soon realized was Dinétah and felt comforted by feelings of home. But these feelings vanished when, "from beneath the sand surfaced a large black creature whose scales moved quickly . . . It was a large snake of monumental proportions . . . I saw not just one creature, but several more emerging from the desert floor. Multiple shadowy heads emerging from the crimson sand." Her dream state connected Diné ancestral homelands with the Lakota imagery of a black snake, linking the dangers of extractivism in one Indigenous community to those of another. Montoya views this vision as a manifestation of kinship and relational ties, and she clarifies: "To seek relationality isn't the performance of a sort of intertribal or even multicultural utopia, but it is about demanding and upholding the Indigenous terms of our existence. We exist because we have maintained enduring relations" (Birkett and Montoya, 2019, pp. 263–64).

It is with this relationality in mind that I have paired the works of Robbins and Trimble-Yanu in this chapter; though they hail from different Indigenous cultures and address concerns specific to those communities, they are powerfully united in their commitment to countering the gendered violence of extractivism.

Notes

[1] It is always preferable to refer to specific community identities–Lakota, for example, narrowing further to Oglala Lakota Sioux Nation (in the case of artist Angelica Trimble-Yanu). Broader terms, however, are sometimes necessary to reference multiple communities at once. In this essay, I have chosen to use "Native" and "Indigenous." These are terms used by several of the scholars cited in this text, and the artists under review also consider this terminology to be appropriate.

[2] The term *Navajo* is of Spanish origin. Members of this community call themselves *Diné*, which translates to *The People*.

[3] A Chapter House is a community center and seat of local governance on the Navajo Nation. The Navajo Nation has 110 local chapters, divided among five regional agencies.

[4] In 2020, Trimble-Yanu produced *Makhá*, a short film that is a continuation of *Iyeska*, including in the documentary photographs and video that the artist produced in 2018. *Makhá* is available at: https://filmsight.com/project/makha/.

References

Allen, P.G. (1986). The Sacred Hoop: Recovering the Feminine in American Indian Traditions. Boston: Beacon Press.

Berlo, J.C. (2009). Alberta Thomas, Navajo Pictorial Arts, and Ecocrisis in Dinétah. In Braddock, A.C. and Irmscher, C. (Eds.), A Keener Perception: Ecocritical Studies in American Art History, pp. 237-253. Tuscaloosa: University of Alabama Press.

Birkett, T.M. and Montoya, T. (2019). For Standing Rock: A Moving Dialogue. In Estes, N. and Dhillon, J. (Eds.), Standing with Standing Rock: Voices from the #NoDAPL Movement, pp. 261-280. Minneapolis: University of Minnesota Press.

Burnette, V. (2018). Confessions of an Iyeska. Salt Lake City, University of Utah Press.

Deer, S. (2015). The Beginning and End of Rape: Confronting Sexual Violence in Native America. Minneapolis: University of Minnesota Press.

Denetdale, J.N. (2007). Reclaiming Diné History: The Legacies of Navajo Chief Manuelito and Juanita. Tucson: University of Arizona Press.

Ellis, E. (2019). Centering Sovereignty: How Standing Rock Changed the Conversation. In Estes, N. and Dhillon, J. (Eds.), Standing with Standing Rock: Voices from the #NoDAPL Movement, pp. 172-197. Minneapolis: University of Minnesota Press.

Gilio-Whitaker, D. (2019). As Long As Grass Grows: The Indigenous Fight For Environmental Justice, from Colonization to Standing Rock. Boston: Beacon Press.

Green, J. (2017). Taking More Account of Indigenous Feminism: An Introduction. In Green, J. (Ed.), Making Space for Indigenous Feminism, pp. 1-20. Halifax: Fernwood Publishing.

Hämäläinen, P. (2019). Lakota America: A New History of Indigenous Power. New Haven: Yale University Press.

Hawley, E.S. (2022a). Native Feminisms and Contemporary Art: Indigeneity, Gender, and Diné Resurgence in the work of Natani Notah and Jolene Nenibah Yazzie. In Hannum, G. and Pyun, K. (Eds.), Expanding the

Parameters of Feminist Artivism, pp. 29-49. New York: Palgrave Macmillan.

Hawley, E.S. (2022b). From the Ground Up: Contemporary Diné Artists Enlist Native Feminisms in a Fight for Environmental Justice. Art in America (November 2022), pp. 66-71.

Johnston, B.R., Dawson, S.E., and Madsen, G.E. (2007). Uranium Mining and Milling: Navajo Experiences in the American Southwest. In Johnston, B.R. (Ed.), *Half-Lives & Half-Truths: Confronting the Radioactive Legacies of the Cold War*, pp. 97-116. Santa Fe: School for Advanced Research Press.

Lame Deer, J.F. and Erdoes, R. (1972). Lame Deer: Seeker of Visions. New York: Pocket Books.

Lyons, S.R. (2000). Rhetorical Sovereignty: What Do American Indians Want from Writing? College Composition and Communication, 51(3), pp. 447-468.

M'Closkey, K. (1998). Weaving and Mothering: Reframing Navajo Weaving as Recursive Manifestations of K'e'. In Ilcan, S. and Phillips, L. (Eds.), Transgressing Borders: Critical Perspectives on Gender, Household, and Culture, pp 115-127. Westport, CT: Bergin & Garvey.

Mihesuah, D.A. (2003). Indigenous American Women: Decolonization, Empowerment, Activism. Lincoln: University of Nebraska Press.

Nagle, M.K. and Steinem, G. (2016). Sexual Assault on the Pipeline, Boston Globe, 29 September. Available at: https://www.bostonglobe.com/opinion/2016/09/29/sexual-assault pipeline/3jQscLWRcmD12cfefQTNsL/story.html

Navajo Nation Environmental Protection Agency (n.d.,a). Home. Available at: https://navajoepa.org/

Navajo Nation Environmental Protection Agency (n.d.,b). What You Should Know About Radon [pamphlet].

Navajo Water Project (n.d.). About the project. Available at: https://www.navajowaterproject.org/project-specifics

Penman, S. (2000). Honor the Grandmothers: Dakota and Lakota Women Tell Their Stories. Saint Paul: Minnesota Historical Society Press.

Powers, M. N. (1988). Oglala Women: Myth, Ritual, Reality. Chicago: University of Chicago Press.

Reyer, C. (1991). Cante ohitika Win (Brave-hearted Women): Images of Lakota Women from the Pine Ridge Reservation South Dakota. Vermillion: University of South Dakota Press.

Robbins, E. (2021). "5,712 + Thirty Percent: An Artist Talk with Emma Robbins", Artist Talk presented at: Native Feminisms, apexart, 6 February.

Robbins, E. (2022). Artist Talk, Santa Clara University, 24 February.

Smith, A. (2003). Not an Indian Tradition: The Sexual Colonization of Native Peoples, Hypatia, 18(2), pp.70-85.

St. Denis, V. (2017). Feminism Is for Everybody: Aboriginal Women, Feminism and Diversity. In Green, J. (Ed.), Making Space for Indigenous Feminism, pp. 42-62. Halifax: Fernwood Publishing.

Szasz, M.C. (1994). Introduction. In Szasz, M.C. (Ed.), Between Indian and White Worlds: The Culture Broker, pp. 3-20. Norman: University of Oklahoma Press.

TallBear, K. (2019). Badass Indigenous Women Caretake Relations: #Standingrock, #idlenomore, #blacklivesmatter. In Estes, N. and Dhillon, J. (Eds.), Standing with Standing Rock: Voices from the #NoDAPL Movement, pp. 13-18. Minneapolis: University of Minnesota Press.

Trimble, C. (2012). Iyeska. Indianapolis: Dog Ear Publishing.

Trimble-Yanu, A. (2022). "Reflections on Returning" Artist Talk presented at: Institute of Contemporary Art San Francisco, 9 December.

Tsosie, R. (1988). Changing Women: The Cross-Currents of American Indian Feminine Identity, American Indian Culture and Research Journal, 12(1), pp. 1-37.

Voyles, T.B. (2015). Wastelanding: Legacies of Uranium Mining in Navajo Country. Minneapolis: University of Minnesota Press.

Walker, J.R. (1991). Wohpe and the Gift of the Pipe. In DeMaille, R.J. and Jahner, E. A. (Eds.), Lakota Belief and Ritual, pp. 109-112. Lincoln: University of Nebraska Press.

Chapter 7
Water, Women, and Resilience in the Andes of Peru

Nicole Sault

For peoples of the Andes, life emerges from water and flows to connect landscapes and all beings, from llamas to condors. In struggling to protect sacred bodies of water, Indigenous communities and women in particular are facing increasing danger. This chapter addresses the centrality of water and Peruvian women's resilience in confronting extractive processes that both consume and contaminate water, obstructing the flow upon which life depends.

Water Connections with Animals, Mountains, and Stars

The Quechua and Aymara speaking cultures of the Andes understand the world as a system in continuous flow that regenerates life through reciprocal exchanges that connect the underworld, middle world, and upper world (Sault, 2021). The relationship with water is understood as reaching up to the mountains where rain clouds gather, flowing back down to plants and animals, and then rising back up to the stars. What we call the Milky Way was known to the Inca as Mayu in Quechua, the great celestial river of shining bright stars, and the spaces in between were known as *hana phuyu* or "dark stars" (Urton, 1981). In this aquatic system, water descends from the stars down to earth, sent by the Mother Llama star as rain or snow, and is drawn back up through the sea, clouds, and fog, which are the earth's breath.

The depth, breadth, and intimacy of the relationship to water is immeasurable, for through water, relationships of kinship are expressed with not only the ancestors and descendants but with one's surroundings, connecting people to an animate world. Fish and flamingos have obvious watery associations that are ritually elaborated, but also condors and bears who protect the snowy mountains. Seashells and seaweed are used in ceremonies not only along the coast but in the highlands as well (Gose, 1994, p. 209; Reinhard, 1985).

These interrelationships demonstrate the intimate connection people have with their surroundings *(entorno)* that is understood as a relationship of kinship extending to water and mountains—"the landscape is not just sacred, but *living*" (Doyon, 2006, p. 352). There are annual ceremonies for cleaning the irrigation canals, not only for practical reasons but also as a sacred duty. Water is a sentient being, flowing down from the heavens and the mountains, and is guarded by condors who are senior kin with authority to punish the disrespectful (Sault, 2018, p. 35).

Figure 7.1. *Terraces with irrigation canals. Colca Canyon, Peru, 2011. Photo by Nicole Sault.*

The aquifers themselves express kinship relations, for according to Inca origin myths, these subterranean waterways were traveled by the ancestors, who emerged at certain locations to establish the homelands where their descendants now live (Doyon, 2006, p. 369). In the contemporary highland community of Puquio, these underground waterways are said to be "the blood flowing through the veins of these Lords" of the mountains (Arguedas 1956, p. 200–201, cited in Doyon, 2006, p. 365).

Kinship understandings also apply to the relationships between different bodies of water. As scholars have observed, in Andean cultures water can be either feminine or masculine and exist in pairs as complementary oppositions. Female waters are circular pools that include lakes, springs, and the womb. Male waters are channeled through rivers, irrigation canals, poured liquids, and semen (Doyon, 2006, p. 354; Salomon, 2017, p. 10). Both mountains and lakes can be pairs of male and female and can marry or be related as brothers and sisters (Sánchez Garrafa, 2006, p.161; Doyon, 2006, p. 368).

In Quechua or Aymara speaking communities of Peru, both men and women identify with water and feel bound to guard and protect the rivers, lakes, springs, glaciers, and other water beings. In water ceremonies women and men may have different roles, which is also true for defending water. In these societies "water does not exist as an isolated thing but is woven into the fabric" of life through liquid cycles in which people and their deities participate (Sault, 2021).

While water brings life, water can also take life through snowstorms, avalanches, flooding, and drought, which underscores the importance of respecting water and observing the rules for proper behavior. When these rules are ignored, there can be consequences, as the mountain deities can send downpours that cause flooding or withhold the rain and cause drought. This also means that assaults on mountains and water are believed to have spiritual reverberations that affect everyone. The land has agency and can respond.

The Politics of Water Extraction

Changes in the climate began to take their toll on Peru decades ago. This has been disastrous for key crops, such as potatoes and corn. Sudden heavy rains lead to potato blight, while unseasonal hail destroys the potato flowers to such an extent that farmers are losing native varieties of potatoes handed down by their ancestors (CIP, 2019). Rising temperatures are melting the mountain glaciers. As the snow disappears, the streams dry up and the pastures and fields suffer. Snow on mountains is seen as a protective veil, so if the snow disappears, the mountain is weakened (Millones and Mayer, 2012). Of all the nations of South America, Peru is the "most water-stressed country" (Bebbington and Williams, 2008, p. 191). Highland pastures as well as crops suffer from drought. In the 1980s and 1990s drought "spurred increased migration from rural areas to cities in Peru" (Fraser, 2009). The number of "water refugees" in the Andes continues to grow, as water disappears and people are forced to abandon their communities to look elsewhere to survive (Carroll and Schipani, 2009). Meanwhile, monocrop export agriculture expands and the populations of cities swell, intensifying stresses on the water system. Dams and drinking water companies are adding to the scarcity (Boelens and de Vos, 2006).

Figure 7.2. *Bare mountain whose protective veil of snow has melted. Arequipa, Peru, 2014. Photo by Nicole Sault.*

The Peruvian government's water policies and water concessions to mining companies have swept aside traditional agricultural practices for allocating water in highland Indigenous communities and usurped the authority of local leaders. The distribution of water for irrigation canals was traditionally supervised by each community's "water judges" or *unu kamayoq* in Quechua who ensured that customary law was followed (Gose, 1994, p. 97; Salomon, 2017, p. 14). However, "state policies and legislation often do not respect historical and local rights systems. Moreover, the decision-making power of the state water bureaucracy is often based on undemocratic principles and under-representation of local communities. Especially powerful economic actors therefore tend to benefit from current privatization and deregulation policies" (Boelens and de Vos, 2006).

Within this context of water loss, scarcity, and uncertainty, mining has brought increasing threats due to contamination, depletion of water supplies, and the disappearance of lakes and springs (Perreault, 2013; Li, 2015). Other consequences of mining include deforestation, depletion of fauna, erosion, sedimentation, salinization, and desertification (Sanchez Soto, 2014). Mining depletes and contaminates water in many ways, diverting water to process ore and to maintain the mining camps of migrants from around the country, desperate to find work. Melting glaciers have further exposed mountains and high valleys, enabling mining projects to move ever higher. The highest permanent settlement in the world is La Rinconada, a gold mining camp above 17,000 feet (Dickerman and Delano, 2019). As Julia Cuadros of the Citizens Movement Against Climate Change explains: "For the open pit mines in the Andes, they cut off the tops of the mountains. The clouds that once rested and released some of their water, now pass over to the other side of the mountain. This produces changes in the pattern of wind and rain, altering the renewal of the aquifers. This local change in climate shows us what will be global" (Banchón, 2015). Yet the Peruvian government's response has been to suggest building pipelines to channel water from higher in the Andes or to reroute rivers that flow to the Amazon (Fraser, 2009).

Processing the ore requires enormous amounts of water, after which the contaminated water is dumped into containment ponds, lakes, or streams and then enters the ground water system and wells, eventually reaching the aquifers. When abandoned mining pits fill with water, the result is acidic lakes with high levels of toxic metals (Bebbington and Williams, 2008). Toxic metals contaminating the water include mercury, lead, cadmium, arsenic, sulphur, iron, zinc, and tin (Garcia et al., 2008; Rojas and Vandecasteele, 2007). The presence of these metals has led to massive die-offs of fish, frogs, and water birds, contaminated crops like potatoes and beans, killed livestock, and damaged the health of communities living near or downstream from the mines (Cabellos, 2015; Gammons et al., 2006; González Pinell, 2011; Rojas and Vandecasteele, 2007; Sánchez, 2015; Chambi Parisaca, Orsag Céspedes, and Niura Zurita, 2012; Garrido et al., 2017; Oporto, Vandecasteele, and Smolders, 2007; Corporate Responsibility Initiative, 2019).

Contamination from mining wastewater is linked to a number of illnesses, including cancer, nosebleeds, birth defects, miscarriages, anemia, arthritis, headaches, neurological dysfunction, paralysis, and metabolic and digestive problems (Garcia et al., 2008; Rojas and Vandecasteele, 2007; Sánchez, 2015; Burgos, 2017). Lead attacks the brain, heart, and kidneys (Burgos, 2017). Accidents also occur, such as the world's largest mercury spill at the Yanacocha Mine in Choropampa, Cajamarca in 2000 (Arana-Zegarra, 2009). Uranium mining is expanding to at least three sites in Peru, which greatly exacerbates environmental and health concerns (Paredes, 2014).

Corporate Rights versus Human Rights

The spread of mining operations throughout the country has restructured the economy and transformed the political organization of Peru. One measure of this is the annual international mining conference organized by Perumin and held each September in Arequipa, southern Peru (Perumin, 2017). The predominant mining companies attending are from Canada, the United States, and China. Since 1990 Peruvian water law has sought to impose a "water culture" that replaces an Indigenous collective understanding of water stewardship with an individualistic mercantile model that promotes government control and enables taxation (Paerregaard, Stensrud, and Andersen, 2016).

Throughout the Americas or *Abya Yala*, there are tremendous problems with megamining projects, but the intensity of mining conflicts is greatest in Peru particularly regarding the usurpation of land and water in Indigenous and other *campesino* communities (Sault, 2018, p. 33). As Li has noted: "While other Latin American countries, from Ecuador to Guatemala, also saw a rise in mining conflicts in the 1990s and 2000s, the intensity and frequency of conflicts in Peru was unparalleled" (Li, 2015). Over half of Peru's six thousand farming communities are torn by mining conflicts (DeEchave and Torres, 2005, cited in Li, 2015).

The government promises consultation with local communities before mining projects are begun. However, these promises often go unfulfilled, and in many cases when a community opposes a mining project overwhelmingly, their objections are ignored. For example, after the 2007 referendum in the northern Peruvian community of Tambogrande, Piura, about the Canadian project of Manhattan Minerals, despite 93 percent of the residents voting against the project, "the company, the central government, and the President of Peru continue to insist that the mine go ahead" (Bebbington and Williams, 2008, p. 192). The situation has been exacerbated by the Peruvian government giving unlimited water concessions in perpetuity to 248 mining companies working in drought-stricken areas (Conflictos Mineros, 2018). These mining leases are also used to justify land seizures and the forced relocation of villages (CNN Español, 2016; Li, 2015; Perreault, 2013; SERVINDI, 2007). By arguing that they will bring in jobs, multinational corporations justify the plunder of the land, water, and people, exploiting poverty.

Big money is being made by international mining companies, and such vast amounts also encourage extensive corruption through buying off government officials to support mining projects and alter environmental reports (El Comercio, 2015). As a consequence, local communities are ignored or lied to when they complain and demand that their rights be honored. In Apurimac when the environmental report for the Las Bambas mine was altered, protests resulted, during which 23 people were injured, three were killed, and many were arrested (Coordinadora Nacional de Derechos Humanos, 2015).

In response to community protests, the mining companies hire Peruvian special forces police from the Dinoes (Dirección Nacional de Operaciones Especiales) as private security, to intimidate and threaten people into forcing them off their land (CNN Español, 2016; Sampat, 2014; SERVINDI, 2007). These private security guards are also used to prevent protestors from blocking access roads and to keep them away from mining sites and equipment. The Peruvian military is also called in as backup (Deutsche Welle, 2015). Clashes between protesters, the police (and soldiers) have led to many injuries and deaths of *campesinos* in the regions of Apurimac, Cusco, Cajamarca, and Arequipa (Rojas, 2015; Deutsche Welle 2015; Li, 2015; Sampat, 2014).

During confrontations, the police have total impunity in using violence against protesters, as the Peruvian government passed Law 30151 in 2014, stating that "Members of the armed forces and the National Police" cannot be held accountable or punished for injuring or killing protesters—they are "exempt from criminal responsibility" while on duty (Dearden, 2014). Since the 2009 attack against 2,000 Aguaruna and Wampi Indigenous People of Bagua in the northern Amazon, human rights defenders and environmentalists fear that the police have been given "a license to kill" (Dearden, 2014). There are also reports of "intimidation, death threats, surveillance and 'judicial harassment'" (Dearden, 2014). Government officials dismiss the protesters' concerns by referring to them as vandals and *delincuentes* (delinquents) in order to demonize them and justify repression.

Human rights organizations such as Amnesty International, Oxfam International, Human Rights Without Borders-Cusco (DHSF), Front Line Defenders, Campaña Minería Perú (Kampagne Bergwerk Perú), and the United Nations have decried the spreading violence associated with mining in Peru (Dearden, 2014; Rojas, 2015; Wagner, 2016). Abuses occurring in mining camps include: forced labor, child labor, prostitution, human trafficking, drug cartels, money laundering, and land theft (Hill, 2016; Wagner, 2016). A report by the Global Initiative Against Transnational Organized Crime (GIATOC), states that in southeastern Peru, in the Department of Madre de Dios, "thousands of girls, some as young as twelve, are being sexually exploited by the miners" (Hill, 2016).

In 2020 over a third of the murders of environmentalists committed worldwide were committed against Indigenous communities, and three out of every four of these deaths occurred in Latin America (Paz Cardona, 2021).

When communities protest against the widespread violence and impunity, environmental destruction, and bodily harm the government response is repression. Media reporting with statistics on the numbers of dead, injured, tortured, and displaced people, usually does not distinguish women from men, even though women may be targeted because of perceived vulnerability or for the opposite—because of their leadership roles. Yet there are several sources of hope for justice—including community defense traditions, women water defenders, and the recognition of "the rights of water." The next sections highlight the leadership roles that Indigenous women of Peru have taken in resisting oppressive methods for extracting ore, water, and human labor.

Reclaiming Community Defense Traditions in Peru

The combination of government officials promoting and protecting mining corporations, together with the cutbacks on environmental regulations, failure to enforce the laws, and government repression have led to mounting frustration and mistrust in communities throughout Peru and other Andean nations, including Bolivia, Chile, and Ecuador. A Water Law and Indigenous Rights Program (WALIR) was formed in 2002 with partners in Bolivia, Chile, Ecuador, Peru, Mexico, France, the Netherlands, and the United States to promote justice in water policies and legislation in South America, but enforcement and protection remains a concern (Boelens and de Vos, 2006). Water Laws have been enacted in Ecuador and Chile (1993), Bolivia (2004, 2010), and Peru (2009), but competing interests prevented water protections from being implemented among Indigenous Peoples and other *campesino* or farming communities. For example, nations like Venezuela that enact restrictions against rampant mining are accused by corporate elites in Canada and the United States of "resource nationalism," hoarding the minerals for their own nation's use (Paez Victor, 2022).

Zibechi noted that: "when states act to facilitate the business of multinationals and leave communities unprotected, as with mining, those communities have no choice but to defend themselves by their own means—through self-defense organizations, mobilization of affected communities, or the creation of new ways to prevent dispossession" (Zibechi, 2014). In South America, Central America, and Mexico, Indigenous Peoples, along with other *campesinos*, are turning to traditional methods of community defense and conflict resolution. Indigenous communities in Peru understand the right to water as inherited from the ancestors and the deities, so the usurpation of these water rights denies a community's connection to the land, their ancestors, and their culture. To protect against the ongoing violence and theft of land and water, they are calling upon their cultural traditions and reinstating ancestral practices.

Figure 7.3. *Inca ceremonial fountain. Ollantaytambo, Peru, 2011. Photo by Nicole Sault.*

In rural northern Peru, *campesino* patrols began in the 1970s to stop cattle theft but decades later these patrols have been adapted for a different purpose—to address contemporary conditions regarding mining. The patrols are based on a justice system that is "managed collectively by the assembly" and emphasizes education and reconciliation over punishment, operating "with relative autonomy from the state" (Zibechi, 2014). These *campesino* patrols focus on protecting the commons and creating alternative development models and programs, preparing prior consultation meetings, providing information and security during referenda, and organizing protests. By 2012 they were being called Guardians of the Lagoons and were organized to regulate access to communal land and water by installing gates and permanent camps for the guardians. In 2014, the Guardians of the Lagoons in northern Peru proposed strikes against projects like the Conga megamine and were successful in slowing down or stopping some of the mining projects (Zibechi, 2014).

Due to protests and strikes organized by various groups, including water guardians, the second largest mine in Peru, Las Bambas, has been forced to shut down temporarily, which has affected stock prices of the company (Cossins-Smith, 2023). As the Las Bambas copper mine was set to expand, protests also mounted over low pay, dangerous working conditions and the ouster of President Carlos Castillo in December of 2022. Despite a truce, the blockades of the supply routes to Las Bambas and two other mines are set to continue (Cossins-Smith, 2023).

Peruvian defenders of water and land view spiritual connections as integral to their ability to engage and sustain their efforts over time. For example, they are drawing upon Inca traditions that recognized stones as living beings who could become animate and were given offerings. These traditions are still practiced, as when stone boundary markers are given libations and can be "invoked in land occupations or border disputes" (Gose, 1994, p. 241). Such practices show how symbolic rituals are used for organizing "the social structure in relation to the local ecology" (Sault, 1998–99, p. 152). In negotiating with government representatives, Andean communities in Peru, Bolivia, and Colombia have also responded with "creative reform proposals to secure Indigenous water rights and water system management" by adopting a strategy of *protesta con propuesta* or "protest and proposal" as alternatives to the government plans for mining industries (Boelens and de Vos, 2006). Women's Patrols are spreading and becoming some of "the principal protagonists in territorial defense," even in areas considered the most patriarchal (Hoetmer, cited in Zibechi, 2014).

Women Water Guardians in Peru

Among Indigenous communities in Peru, women's relationships with water have a depth and complexity that goes beyond Western categories of nature and politics (De la Cadena, 2010). Indigenous and other *campesino* leaders who are defending water include both women and men, in recognition of the responsibility that both women and men have for protecting and nurturing mountain lakes, rivers and water cycles. Women are often the leaders who organize resistance, their sense of solidarity and activism arising from their spiritual relationships to water as well as concern for their families (Gudynas, 2022). This concern for the well-being of others extends to their flocks of camelids and fields of potatoes and quinoa as well as other beings, including birds, fish, and mountains.

A gendered understanding of water means that the threats posed by extractive mining policies against land and water are perceived somewhat differently by women and men, according to the body of water that is endangered. Women have essential relationships to lakes and lagoons that are threatened by mining operations. Men have essential relationships to the rivers that flow into and out of lakes and lagoons, which are likewise threatened by mining. Women as well as men are responsible for protecting water in various forms, and thus both men and women organize as water protectors who are recognized and honored by their communities. While men patrol at night, both women and men patrol during the daytime, and women have taken on central roles in organizing groups of women to protect their homes and the lagoons.

Jenkins notes that Andean women water guardians describe their role as protecting a mother who is the source of life (Jenkins, 2014). Their connection to water includes "the *first water* of the womb, and water as a living being who they are called to honor and protect as part of their heritage and identity" (Sault, 2018, p. 36). Women's role as guardians of lagoons grows out of their particular association with pooling water and their sacred connection to *Mamacocha* (Mother Water).

In the Andes, women's roles in resisting mining megaprojects have frequently gone unrecognized, due to patriarchal colonial assumptions that men are leaders and risk-takers with authority, while women are presumed to be supporters or passive victims. However, Indigenous women have resisted colonization from the beginning, and in northern Peru they have demonstrated key leadership roles in direct action against extractive mining and the ensuing environmental destruction along with political repression. They have developed creative strategies for defending themselves, their families, and their communities (Jenkins, 2014).

Women water protectors who have gained international recognition include Máxima Acuña Atalaya de Chaupe of Cajamarca in northern Peru. In 2016, she was awarded the Goldman environmental prize, based on her struggle to protect the *Laguna Azul* (Blue Lagoon) against the Conga mining company from the United States (CNN Español, 2016). This mining company was trying to force Maxima off her land in order to use the area as a dumping ground for contaminated waste water. She was awarded the Goldman prize for her courageous resistance, despite the ongoing violent attacks by the mining company's security forces and the Peruvian military who beat her and her daughter unconscious and killed their dog (CNN Español, 2016).

Another undaunted woman water protector from the Cajamarca region is Nélida Ayay Chilón, a Quechua woman who became a lawyer in order to better defend the lagoon she is protecting, Laguna Mamayacu. In the documentary *Hija de la Laguna* (*Daughter of the Lake*), Ayay explains that the lagoon is her mother to whom she offers flowers and photos of the five farmers who were killed in 2012 for defending the lagoon. She says to the lagoon "only by looking after you well can you continue providing for us always" (Cabellos, 2015). She has been called the voice of the sacred mountains and the animals, "the heart of the defense of the land and the water in Cajamarca" (Muñoz, 2018). The repression by the Yanacocha mining company is ongoing, but as Nélida Ayay explains, there are many women of water who are resisting and we should be thankful for them (Muñoz, 2018).

Women water defenders in Peru have been assaulted, beaten, raped, and killed, yet these women continue to lead the protection of the land and water of their communities. Despite the toll of physical violence and threats, women explain that they find strength from their connection to the spiritual traditions that flow from the ancestors. One specific way they access these connections is through dreams that inspire their songs. In the video, *Singing, Dreaming and Resistance Among the Water Guardians of Cajamarca, Peru*, women talk about singing to raise up their voices and show courage to inspire other women. During their struggles to defend themselves and their homes against the Conga gold mine megaproject, both Máxima Acuña Atalaya de Chaupe and her neighbor, Señora Santos describe the importance of the dream-inspired songs they sing (Santiago, 2022).

In southern Peru, two Quechua-speaking sisters are reclaiming ancestral traditions of *Qucha ruway* or "sowing and harvesting water" to protect sacred lagoons and address the problem of drought (Rodríguez, 2020b; Euroclima, 2020). Magdalena and Marcela Machaca Mendieta are agricultural engineers from Ayacucho who are building reservoirs and canals high in the mountains. Following their grandfather's teachings from the 1970s, they are "nurturing water" using traditional techniques along with ceremonies and songs. Since 1995 they have built 120 reservoirs around their community, sealing leaks with soil and using special plants to protect the soil, filter the water, and shelter birds (Rodriguez, 2020a). These reservoirs are recharging the aquifers and ground water, renewing springs that had previously gone dry.

Magdalena says that they sing in Quechua to express their feelings to the water (Euroclima, 2020). They describe the lagoon as a living being and emphasize that "we should never stop showing her affection and understanding" (Rodríguez, 2020a; Euroclima, 2020). For the two sisters "the lagoons play the role that the frozen mountain-tops used to play" in providing water. As they explain, "our communities are the protectors of water and we are proud of that" (Rodríguez, 2020a). They founded an organization called *Asociación Bartolomé Arypalla* (ABA), to promote the "sowing of water" and have traveled to Piura in northern Peru, as well as other countries like Costa Rica, to teach these techniques (Euroclima, 2020).

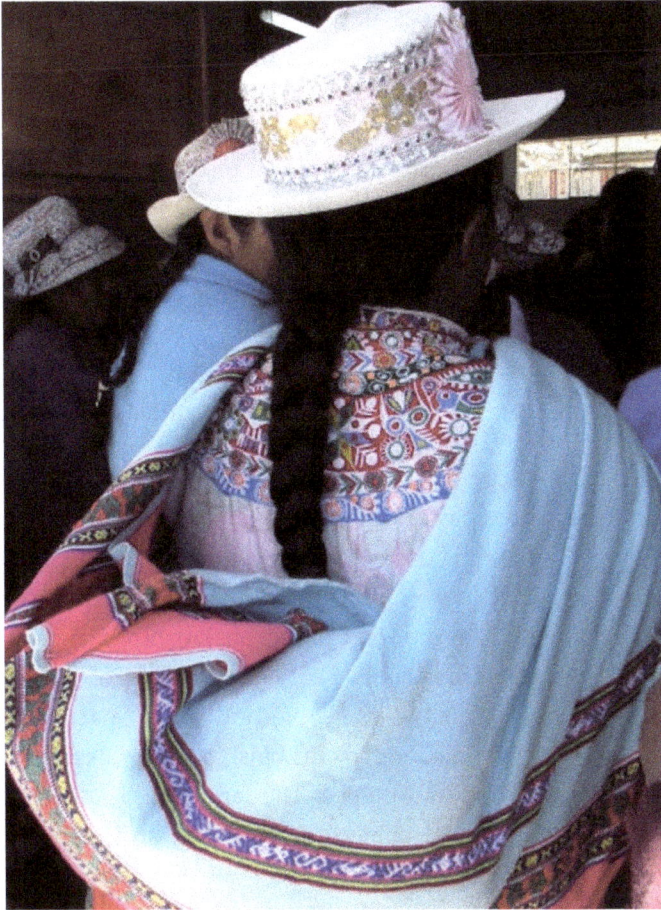

Figure 7.4. *Peruvian woman wearing a hat that represents a guardian mountain, and a blouse embroidered with potato flowers and birds. Colca Canyon, Peru, 2013. Photo by Nicole Sault.*

Water Laws and Water Rights

How we relate to our surroundings is expressed in our relationship to water—whether as a resource to be extracted for profit or as a living being who is respected and honored as sacred. Porcel Bugueño notes that in Latin America both the educational system and work training programs are directed toward the goal of extractivism (Porcel Bugueño, 2021, p. 136). The capitalist language used to describe water commodification and extraction is so internalized that it goes unrecognized or is lauded as the only perspective worthy of consideration in promoting "progress."

International development agencies and Peruvian officials who promote mining and dam projects portray water as a resource that should be used "rationally" and extracted for financial gain. Water is reduced to units of measurements in cost-benefit analyses, according to which water must be "used" or it is "wasted." This perspective portrays nature as a commodity for sale to the highest bidder, representing a worldview of alienation or estrangement from the water and land.

In conflicts over water and mining, the language of commodification is used to advance a capitalist extractive ideology and suppress other worldviews that are based on collective access by communities. Commodity terminology is used to justify extractive processes and deny the social and spiritual meaning attributed to water in cultures that still maintain a relationship to the land based on kinship, stewardship, and reciprocity. The emphasis on water commodification obscures the underlying relationships between people and water, particularly when these relationships are embedded in kinship networks that connect communities to ancestors and deities. The language of commodification legitimizes "deeply political choices that protect and stabilize specific political orders" (Boelens et al., 2016, p. 3). In Cochabamba, Bolivia, a coalition of Indigenous "sowers of water" from throughout the Americas created the Declaration of K'oari, which observes that: "Under capitalism, the administrative institutions of water only are concerned with the distribution of water; as a resource which is measured, allocated and sold" (Declaration of K'oari, 2018).

In the struggle for control over water, the ability to culturally define and politically organize water rights and territories is central.[1] Regarding water rights and forms of governance in the Andes, Boelens argues that "attempts to modernize and normalize users through universalized water culture, 'rational water use' and de-politicized interventions deepen water security problems rather than alleviating them" (Boelens, 2015). He notes that "user collectives challenge imposed water rights and identities, constructing new ones to strategically acquire water control autonomy and re-moralize their waterscapes" (Boelens, 2015). What has encouraged this call to "re-moralize their waterscapes" is the recognition of the spiritual connection to water as central to Indigenous religious traditions, political structures, and economic organization in the Peruvian Andes.

Throughout the Americas and the wider world, there have been calls for the recognition of water rights in alignment with communal models of access, rights and responsibilities that grow out of ancestral spiritual connections to land and water. In the Andes, these rights and responsibilities are recognized by Quechua and Aymara speakers in relation to *Pachamama*, the Earth Mother. McGoldrick refers to *Pachamama* in his discussion on the declaration of the Rights of Nature, first passed in Ecuador's constitution of 2008. This concept was later taken up in the 2010 rewriting of the Bolivian constitution, the first in the world to include the "Law of the rights of the earth" which protects the earth and all living systems as legal persons (McGoldrick, 2017, p. 476). The law "states that water can never be privatized" and people have a right to sue when water rights are infringed upon (McGoldrick, 2017, p. 470). McGoldrick describes the recognition of solidarity with *Pachamama*, the Earth Mother, as "a connection and vision of nature as a sacred whole that the human community itself is contained within" and that this "makes it impossible to abstract and instrumentalize water or nature as a commodity" (McGoldrick, 2017, p. 477).

Living in solidarity with one's surroundings (*entorno*), *Sumak Kawsay* is "an Indigenous principle promoting living in harmony with nature in order to achieve wellbeing for all" (Pachamama Alliance, 2023). In Peru the central values for both Quechua and Aymara language traditions emphasize generosity, reciprocity and interdependence. This world view of interconnectedness recognizes that a community's survival depends on how we treat others, both human and beyond, to maintain equilibrium.

Resilience

What I have described above for Peru holds true for much of the Andes, including Bolivia, Chile, and Ecuador, where the centrality of water as life is particularly elaborated (Sault, 2018, p. 36). Water flows, permeates, and transforms, embodying the essence of key cultural values—connection, cycle, reciprocity, return, and renewal. Mining and other extractive industries threaten the land and water because they "interrupt the flow of life, rupture the cycle of renewal, alienate people from other beings, offend the deities, and dishonor the ancestors" (Sault, 2021). The depth of this danger is enormous, because water is about relationships and connections above all else, while extractive industries destroy relationships and break connections, transforming a sacred, living landscape into a collection of resources to be broken and removed to distant lands. Much of Peru is being turned into a "sacrifice zone" in which people and other living beings are contaminated or destroyed (Hedges and Sacco, 2012).

Indeed, mining in Peru has taken on even greater urgency with the discovery of large deposits of lithium that are said to be greater than known reserves in Bolivia, Chile, and Argentina (Wadhwa, 2023). The lithium reserves are being privatized, exploited by a Canadian company, Macusani Yellowcake (owned by Plateau Energy Metals, a subsidiary of American Lithium, that is also mining uranium). Protests against the ousting of President Pedro Castillo and the lithium exploitation have been met with government repression. In May 2023, the U.S. Southern Command was authorized by the Peruvian government to send 700 marines to provide military training for the Peruvian special forces and the police (SwissInfo 2023; Van Auken 2023).

The Colombian inalienable territories called *resguardos* are continually being invaded by outsiders who seize the land for mining, petroleum, logging, ranching, growing coca, and other illegal activities (Montero 2022). In Peru, the destruction that has been wrought is massive, and many communities-have been forcibly relocated or driven out by environmental calamities.

Nevertheless, some Indigenous communities have been successful in restraining mining projects or halting them altogether, including the Conga and Las Bambas mines in Peru and in the Cauca region of Colombia (Zibechi, 2014; Cossins-Smith, 2023). Recognition of the *rights* of water is growing, and Indigenous forms of community justice and environmental protection are being reclaimed and promoted in Peru.

The role of women water guardians demonstrates the power of resilience, even in the face of what seem to be insuperable obstacles. Water teaches resilience and informs their lives. When they perform ceremonies and sing, they are calling on Indigenous traditions to invoke the spirits of the water and the land, invoking their ancestors, and lifting up the names of those who have been killed, along with their photos, not only to honor them but to show that they are present and have ongoing influence (Cabellos, 2015). Through these practices women water guardians demonstrate to their communities that they are not alone, and reassert their connections to the land, the water, and other beings who they call upon for protection. All this safeguards the space around them and encourages others, who witness that women are not voiceless. Their voices ring loud and clear and are lifted up by others in word, ceremony, and song.

Dedication and Acknowledgments

This essay is dedicated to Berta Isabel Cáceres Flores of Honduras, Lenca leader, killed for protecting the Gualcarque River.

This is written with deepest gratitude to the people of Peru who shared their stories with me and trusted that I would lift up their voices. Sadly, I cannot name many of those to whom I am most indebted, due to the risks for people in these mining regions. Those in Peru I can name are Orlando Bedoye, Ursula Podestá, Alejandro Malaga, Eitel Manrique, Enrique Rivera, Cronwell Jara and Cecilia Granadino; and in the United States they are Helen Drachkovitch, Peter C. Reynolds, Sarah Saul, and Ann Marie Saidy.

Notes

[1] For further reading on Water Citizenship in the Andes, Peru's water law, and the imposition of a "water culture" that is taxable, see Paerregaard, Stensrud, and Andersen (2016). The contrast between artisanal mining versus megamining is discussed by Gudynas (2022) and Sánchez Soto (2014).

References

Arana-Zegarra, M. (2009). El Caso de Derrame de Mercurio en Choropampa y los Daños a la Salud en la Población Rural Expuesta. *Revista Peruana de Medicina Experimental y Salud Pública*, 26 (1), pp.113–116. Available at: http://www.scielo.org.pe/scielo.php? script=sci_arttext&pid=S1726-46342009000100019

Banchón, M. (2015). COP21: Minería y cambio climático en América Latina. *Deutsche Welle*, September 12. (Translation mine). Available at: http://www.dw.com/es/cop21-miner%C3%ADa-y-cambio-clim%C3%A1tico-en-am%C3%A9rica-latina/a-18908343

Bebbington, A. and Williams M. (2008). Water and Mining Conflicts in Peru. *Mountain Research and Development*, 28 (3-4), pp.190–195. Available at: DOI:10.1659/mrd.1039

Boelens, R. (2015). *Water, Power and Identity: The Cultural Politics of Water in the Andes.* New York: Routledge.

Boelens, R., Hoogesteger J., Swyngedouw E., Vos J. and Wester, P. (2016). Hydrosocial Territories: A Political Ecology Perspective. *Water International*, 41(1), pp.1-14. Available at: DOI:10.1080/02508060.2016.1134898

Boelens, R. and Vos, H. de (2010). Water Law and Indigenous Rights in the Andes. *Cultural Survival Quarterly*, May 7. Available at: https://www.culturalsurvival.org/publications/cultural-survival-quarterly/water-law-and-indigenous-rights-andes

Burgos, R. (2017). *La Oroya, la Quinta Ciudad más Contaminada del Mundo.* Available at: http://conflictosmineros.org.pe/2017/11/29/la-oroya-la-quinta-ciudad-mas-contaminada-del-mundo/

Cabellos, E. (2015). *Hija de la laguna/ Daughter of the Lake.* A Film. https://cine lasamericas.org/panorama-feature-films-claiff19/2016/hija-de-la-laguna

Carroll, R. and Schipani, A. (2009). Bolivia: water people of Andes face extinction. *The Guardian*, April 24. Available at: https://www.theguardian.com/world/2009/apr/24/andes-tribe-threat-bolivia-climate-change

Chambi Parisaca, L. J., Orsag Céspedes, V. and Niura Zurita, A. (2017). Evaluación de la Presencia de Metales Pesados y Arsénico en Suelos Agrícolas y Cultivos en Tres Micro-cuencas del Municipio de Poopó.

Revista De Investigación E Innovación Agropecuaria Y De Recursos Naturales,
4 (1) 29, pp. 67-93.

CIP International Potato Center (2019). *Potato Guardians reflect on Climate
Change's impact in the Andes.* Available at: https://cipotato.org/blog/potato-
guardians-reflect-climate-change-impact-andes/

CNN Español (2016). *Máxima Acuña, la Campesina que se Enfrentó a una
Multinacional y Salvó el Páramo,* 20 abril. Available at:
http://cnnespanol.cnn.com/2016/04/20/maxima-acuna-la-campesina-que-
se-enfrento-a-una-multinacional-y-salvo-el-paramo/

Conflictos Mineros (2018). *Investigación Encuentra que 248 Mineras Tienen
Licencias de Agua a Perpetuidad,* 2 septiembre. Available at:
http://conflictosmineros.org.pe/2018/02/09/alerta-investigacion-encuentra-
que-248-mineras-tienen-licencias-de-agua-a-perpetuidad/

*Coordinadora Nacional de Derechos Humanos (2015). Ni Una Muerte Más: Un
llamado a la calma y al diálogo en el caso de Las Bambas. 29 septiembre.* Available
at: https://www.business humanrights.org/es/%C3%Ultimas-noticias/un-
llamado-a-la-calma-y-al-di%C3%A1logo-en-el-caso-de-las-bambas/

Corporate Responsibility Initiative (2019). Glencore mine poisons children. *Terre
des hommes schweiz,* November 2. Available at: https://www.terredes
hommesschweiz.ch/en/glencore-mine-poisoned-children/

Cossins-Smith, A. (2023) Protestors to resume blockade of Las Bambas mine in
Peru. March 6. Available at: https://www.mining-technology.com/
news/protestors-resume-las-bambas-blockade/

De La Cadena, M. (2010). Indigenous Cosmopolitics in the Andes: Conceptual
Reflections Beyond "Politics". *Cultural Anthropology* 25(2), pp.334–370.
Available at: DOI:10.1111/j.1548-1360.2010.01061

Dearden, L. (2014). Peruvian police and soldiers given 'licence to kill'
protesters five years after Bagua violence. *The Independent,* June 30.
Available at: https://www.independent.co.uk/news/world/americas/
peruvian-police-and-soldiers-given-licence-to-kill-protesters-five-years-
after-bagua-violence-9574484.html

Declaration of K'oari (2018). Available at: http://ecosocialisthorizons.com/
2018/12/the-declaration-of-koari/

Deutsche Welle (2015). *Tensión en Perú por Huelga Departamental contra Mina,* May 12. Available at: http://www.dw.com/es/ tensi%C3%B3n-en-per%C3%BA-por-huelga-departamental-contra-mina/a-18447007

Dickerman, K. and Whitlow Delano J. (2019). Mining for gold in the world's highest permanent human settlement. *Washington Post,* June 21. Available at: https://www.washingtonpost.com/photography/2019/06/21/mining-gold-worlds-highest-permanent-human-settlement/

Doyon, S. J. (2006). Water, Blood and Semen: Signs of Life and Fertility in Nasca Art. In Isbell W. H. and Silverman H. (Eds.) *Andean Archaeology III: North and South.* New York: Springer.

El Comercio (2015). Las Bambas: ¿Por qué se inició protesta contra proyecto minero? Sept. 28. Available at: http://elcomercio.pe/peru/apurimac/bambas-que-se-inicio-protesta-contra-proyecto-minero- noticia 1844495?flsm=1?ref=nota_peru&ft=mod_leatambien&e=titulo

Euroclima (2020). *Mujeres ayacuchanas preservan el agua frente al cambio climático,* March 7. Available at: https://www.euroclima.org/actualidad/noticia-bosque/696-mujeres-ayacuchanas-preservan-el-agua-frente-al-cambio-climatico

Fraser, B. (2009). Water Wars Come to the Andes. *Scientific American,* May 19. Available at: https://www.scientificamerican.com/article/water-wars-in-the-andes/

Gammons, C. H., Slotton D. G., Gerbrandt, B. Willis Weight W., Young C. A., McNearny, R L., Cámac, E., Calderón,R. and Tapia, H. (2006). Mercury Concentrations of Fish, River Water, and Sediment in the Río Ramis-Lake Titicaca Watershed, Peru. *Science of the Total Environment,* 368(2-3), pp.637–648.

Garcia, M. E., J. Quintanilla, O. Ramos, M. Ormachea, and A. Niura (2008). *Estudio de Contaminación en la Cuenca del Lago Poopó— Bolivia.* Universidad Mayor de San Andrés, La Paz, Bolivia.

Garrido, A. E., Strosnider, W.H.J., Wilson, R.T., Condori, J. and Nairn, R. (2017). Metal-contaminated Potato Crops and Potential Human Health Risk in Bolivian Mining Highlands. *Environmental Geochemistry and Health,* 39 (5641), pp.681–700. Available at: DOI:10.1007/s10653-017-9943-4

Global Data (2023). Peru: Five Largest Mines in 2021. Available at: https://www.globaldata.com/data-insights/mining/peru--five-largest-mines-in-2090833/

González Pinell, B. (2011). Contaminación del Agua en Bolivia. Available at:http://aguabolivia.blogspot.com/2011/02/contaminacion-del-agua-en-bolivia.html

Gose, P. (1994). *Deathly Waters and Hungry Mountains: Agrarian Ritual and Class Formation in an Andean Town.* Toronto, Canada: University of Toronto Press.

Gudynas, E. (2022). Extractivismo en América Latina: ¿Hay caminos hacia el desarrollo sustentable y la justicia socio-ambiental? Seminario Internacional "Extractivismo en América Latina: ¿Hay caminos hacia el desarrollo sustentable y la justicia socio-ambiental?" 19 febrero.

Hedges, C. and Sacco J. (2012). *Days of Destruction, Days of Revolt.* Navayana, India: Nation Books.

Hill, D. (2016). Gold-mining in Peru: Forests Razed, Millions Lost, Virgins Auctioned. *The Guardian.* May 1. Available at: https://www.theguardian.com/environment/andes-to-the- amazon/2016/may/01/gold-mining-in-peru-forests-razed-millions-lost-virgins-auctioned

Jenkins, K. (2014). Unearthing Women's Anti-Mining Activism in the Andes: Pachamama and the "Mad Old Women." *Antipode, 47, pp.* 442–460. *Available at: DOI:* 10.1111/anti.12126

Li, F. (2015). *Unearthing Conflict: Corporate Mining, Activism, and Expertise in Peru.* Durham, North Carolina: Duke University Press.

McGoldrick, T. (2017). The religious case for water as a human right from the Andes. *International Journal of Design and Nature and Ecodynamics,* 12(4). Available at: DOI:10.2495/DNE-V12-N4-470-481

Millones, L. and Mayer, R. (2012). *Fauna Sagrada de Huarochirí,* Lima, Peru: Instituto de Estudios Peruanos.

Muñoz, F. (2018). Nélida Away Chilón, El corazón de la defensa de la tierra y el agua en Cajamarca. *Luchadoras México,* 10 enero. Available at: https://luchadoras.mx/nelida-ayay-chilon-corazon-la-defensa-la-tierra-agua-cajamarca/

Oporto, C., Vandecasteele, C. and Smolders, E. (2007). Elevated Cadmium Concentrations in Potato Tubers Due to Irrigation with River Water Contaminated by Mining in Potosí, Bolivia. *Journal of Environmental Quality*, 36 (4), pp. 1181–1186. Available at: DOI:10.2134/jeq2006.0401

Paerregaard, K., Bredholt Stensrud, A. and Oberborbeck Andersen A. (2016). Water Citizenship: Negotiating Water Rights and Contesting Water Culture in the Peruvian Andes. *Panoramas*, September 16. Available at: https://www.panoramas.pitt.edu/larr/water-citizenship-negotiating-water-rights-and-contesting-water-culture-peruvian-andes

Paez Victor, M. (2022). Canada and the Kidnapping of Ambassador Saab. *Counterpunch*, November 28. Available at: https://www.counterpunch.org/2022/11/28/canada-and-the-kidnapping-of-ambassador-saab/

Paredes, J. (2014). Estudios Revelan Existencia de Uranio en Siete Regiones. *La Razón*, November 21. Available at: https://www.la-razon.com/lr-article/estudios-revelan-existencia-de-uranio-en-siete-regiones/

Paz Cardona, A.J. (2021). Latinoamérica sigue siendo la región más peligrosa para los defensores ambientales. 13 septiembre. Mongobay. Available at: https://es.mongabay.com/2021/09/latinoamerica-asesinatos-defensores-ambientales-global-witness/

Perreault, T. (2013). Dispossession by Accumulation? Mining, Water and the Nature of Enclosure on the Bolivian Altiplano. *Antipode*, 45(5), pp. 1050–1069. Available at:DOI:10.1111/anti.12005

Perumin - Convención Minera (2017). Arequipa, Peru. Available at: https://perumin.com/perumin36/public/es

Porcel Bugueño, J. (2021). Extractivismo en América Latina: Una fresca herida colonial. *Pluralidades*, 7-8, pp. 132-149. Available at: http://pluralidades.casadelcorregidor.pe/pluralidades_7-8/Porcel_Pluralidades-7-8.pdf

Reinhard, J. (1985). Sacred Mountains: An Ethno-Archaeological Study of High Andean Ruins. *Mountain Research and Development*, 5(4), 1985, pp. 299–317. Available at: DOI:10.2307/3673292

Rodriguez, S. (2020a). As glaciers shrink, Peruvian sisters build 'sacred' reservoirs for city water. *Thomson Reuters Foundation*, March 2. Available at: https://www.reuters.com/article/us-climate-change-peru-water-trfn/as-glaciers-shrink-peruvian-sisters-build-sacred-reservoirs-for-city-water-idUSKBN20P10Z

Rodriguez, S. (2020b). Las emprendedoras hermanas Machaca, creadoras de agua donde no hay. *Huchachos.com,* 3 marzo. Available at: https://www.huachos.com/detalle/las-emprendedoras-hermanas-machaca-creadoras-de-agua-donde-no-hay-video-noticia-10189

Rojas, E. (2015). Virulento Rebrote de Conflicto en Perú. *Deutsche Welle.* September 29. Available at:http://www.dw.com/es/virulento-rebrote-de-conflicto-minero-en-per%C3%BA/a-18749889

Rojas, J. C. and Vandecasteele, C. (2007). Influence of Mining Activities in the North of Potosi, Bolivia on the Water Quality of the Chayanta River, and its Consequences. *Environmental Monitoring and Assessment.* 132(1-3), pp. 321–330. Available at: https://pubmed.ncbi.nlm.nih.gov/17171242/

Sampat, P. (2014). Over 80 Groups Sign Statement Opposing Intimidation and Forced Displacement of Mining Protesters in Peru. *Earthworks Action,* April. Available at: https://earthworksaction.org/over_50_groups_sign_statement_opposing_intimidation_and_forced_displacement/#.WYUBqoplBE5

Sánchez, W. (2015). Perú: Confirman contaminación de Aguas por Yanacocha. *Red Latina Sin Fronteras.* Available at: https://redlatinasinfronteras.word press.com/2015/02/10 /peru-confirman-contaminacion-de-aguas-por-yanacocha/

Sánchez Garrafa, R. (2006). *Apus de los Cuatro Suyos:Construcción del Mundo en los Ciclos Mitológicos de las Deidades Montaña.* Doctoral Dissertation, Department of Anthropology, Universidad Nacional Mayor de San Marcos, Lima, Peru. Available at: https://cybertesis.unmsm.edu.pe/handle/20.500.12672/2749

Sánchez Soto, L. (2014). Efectos de la Contaminación Producida por los Relaves de la Minería Informal Puno, Perú. Available at: https://www.monografias.com/trabajos101/efectos-contaminacion-producida-relaves-mineria-informal-region-puno/efectos-contaminacion-producida-relaves-mineria-informal-region-puno

Santiago, C. J. (2022). Singing, Dreaming and Resistance Among the Water Guardians of Cajamarca, Peru. Conference proceedings: *The Seventh Biennial Meeting of the Society of Amazonian and Andean Studies,* March 5-6, James Madison University: Harrisonburg, Virginia. Available at: https://www.academia.edu/video/kGwaJk

Sault, N. (2021). From Sea to Stars— Water Cycles in Highland Peru. Presentation for session on "Water as Life, as Being, and as Person." Society of Ethnobiology Meetings, Cedar City, Utah.

Sault, N. (2018). Condors, Water, and Mining: Heeding Voices from Andean Communities. *Ethnobiology Letters*, 9(1), pp: 27–43. Available at: https://www.researchgate.net/publication/326329888_Condors_Water_and_Mining_Heeding_Voices_from_Andean_Communities

SERVINDI (2007). Perú: Desalojan de sus Tierras a Comuneros de Uyuccasa por Orden de Minera. Available at: https://www.servindi.org/actualidad/2798

Salomon, F. (2017). *At the Mountains' Altar: Anthropology of Religion in an Andean Community.* New York: Routledge.

Swiss Info (2023). *López Obrador dice que sería un "timbre de orgullo" ser considerado "no grato" por Perú.* 22 mayo. Available at: https://www.swissinfo.ch/spa/m%C3%A9xico-per%C3%BA_l%C3%B3pez-obrador-dice-que-ser%C3%ADa-un--timbre-de-orgullo--ser-considerado--no-grato--por-per%C3%BA/48532382

Urton, G. (1981). Animals and Astronomy in the Quechua Universe. *Proceedings of the American Philosophical Society*, 125(2), pp. 110-127. Available at: http://fcaglp.fcaglp.unlp.edu.ar/~sixto/arqueo/curso/Urton%20-%20Animals%20and%20Astronomy%20in%20the%20Quechua%20Universe.pdf

Van Auken, B. (2023). Pentagon sending troops to train Peruvian coup regime's killers. *World Socialist Web Site.* Available at: https://www.wsws.org/en/articles/2023/05/29/nrkx-m29.html

Wadhwa, T. (2023). Peru's coup government is privatizing lithium mining. *People's Dispatch*, April 18. Available at: https://peoplesdispatch.org/2023/04/18/perus-coup-government-is-privatizing-lithium-mining/

Wagner, L. (2016). Organized Crime and Illegally Mined Gold in Latin America. March 30. *Global Initiative Against Transnational Organized Crime.* Available at: https://globalinitiative.net/analysis/organized-crime-and-illegally-mined-gold-in-latin-america/

Zibechi, R. (2014). Community Self-Defense Against Megaminería. *Counterpunch.* July 31. Available at: https://www.counterpunch.org/2014/07/31/community-self-defense-against-megamineria/

Chapter 8

Extractivism in the Country of Good Living: The Dystopian Reality of Black and Indigenous Struggles

Antonia Carcelen-Estrada

What is Life, Anyways?

When I first met Yaku Pérez Guartambel as his translator at NYU in 2013, he was building diasporic coalitions to defend water on a planetary scale. He gifted me his book *Agua u Oro* (2012) which details his defense of Kimsakocha since 2001 against the company IamGold, one of five Canadian mining projects operating in Azuay province. Other Canadian subsidiaries there included Nabominas, Atlas Mineras, Ecuador Gold, and Cornerstone, but there were also British (Wega Minerals) and American (MEC) corporations (Pérez Guartambel, 2012, p. 21–22). This young lawyer was testing out umbrella concepts articulating the new Ecuadorian constitution passed in 2008, like *Sumak Kawsay*, the right to a good living, the human right to water, and the rights of nature. Water or Gold? Through this simple question, Yaku sets his readers to choose between a sustainable future or an extractivist model where the Global North "takes the wealth and leaves misery" (*llevan la riqueza, dejan la pobreza*) (Pérez Guartambel, 2012, p. 22). How can corporations choose gold and pollute water if that directly violates the rights of nature and of Indigenous people's right to self-determination? Or the more basic human right to water? Besides, no one drinks gold.

Kimsakocha is one example of the overlapping Indigenous and environmental movements with competing meanings of life, *Kawsay*, and of the planet, *Pacha*. These differ significantly at every corner of resistance, but movements overlook their differences to form alliances to defend our blue planet. Yaku has so far not only successfully defended Kimsakocha's 32,042 acres of tundra where water feeds many towns downstream, mostly Indigenous communities that will suffer the brunt of extractivist mining, but also the rivers passing through cities like Tarqui and Cuenca (Pérez Guartambel, 2012, p. 21). Communities resist and defend water and nature from this colonial form of development that sometimes takes the euphemistic name of sustainable mining. Yaku used Ecuador's 2008 Constitution of Montecristi to force a legal debate between discrepant extractivist Mining Laws and the Constitutional Rights of Nature: as rights-bearing subject (available online at Organization of American States, 2011, Art. 10); the place where life reproduces (OAS, Art. 71); and deserving restoration of damaged ecosystems (OAS, Art. 72, 73). The Constitution also guarantees water as a human right (OAS, Art. 12), an inalienable public good (OAS, Art. 318) that the government can manage only to protect ecosystems (OAS, Art. 411, 412). Yaku also used international law, like the Dublin Declaration on Water and Sustainable Development (1992), and UN and other scientific reports as evidence in his support. His organizing was local at first, later regional to the Andes, but soon became transcontinental when people from all walks of life gathered in the *Continental Encounter of the Peoples of Abya Yala in Defense of Water and the Pachamama* (2011), where they reached a consensus: *Sumak Kawsay*, a dignified, integral indigenous living—which includes the right to drinking water, irrigation, and clean rivers—was incompatible with an extractivist development that could not be sustainable.

Since that day a decade ago, I have followed Yaku's accomplishments at the national and international levels to advance Indigenous rights and the rights of nature and of water, setting precedents that legal scholars will surely study for decades to come. Yet, the mining struggle is hardly unique to Kimsakocha. Azuay's mining map shows projects surrounding the protected *Parque Nacional Cajas*, and they reveal other signposts to a dystopian future foretold by the lived reality of Indigenous peoples in Canada similarly surrounded by extractivist activities that transform their livelihood, but who still lack access to the most basic goods such as running water, electricity, or sewage (Pérez Guartambel, 2012, p. 29). Yaku compares Kimsakocha to mining projects in Peru, like Oroya's fight against

Southern, which uses 250 liters of water per second, or the almost 18,000 liters of water per day depleted in Yanacocha in the region of Cajamarca, a region now made even poorer after mining began (Pérez Guartambel, 2012, p. 18). Yanacocha, the namesake extractivist project-company, is a subsidiary of Newmont, which also has subsidiary companies in Mexico, alongside Barrick Gold, Goldcorp, or the Golden Group. The latter also operates "without law or authority" (*sin ley ni autoridad*) in Honduras, Guatemala, and El Salvador, creating mining problems that run across national borders (Pérez Guartambel, 2012, p. 20). In Chile, mining increased 150 percent in the first decade of this century and in Argentina the mining agenda promised 10,000 jobs, but could barely deliver a tenth of that, and used mostly imported labor (Pérez Guartambel, 2012, p. 24). Mining does not make economic sense for the local population and its pollution spreads heavy metals such as copper, cadmium, lead, mercury, zinc, nickel, and cyanides (Pérez Guartambel, 2012, p. 31). Water pollution joins gas emissions, deforestation, and soil degradation as some of the effects that Indigenous populations must face directly because of an uneven development that only contributes to widening the global wealth gap (Pérez Guartambel, 2012, p. 35).

The blatant denial of Indigenous constitutional and international rights (i.e., to previous consultation and consent), are common tools used for an extractivist development that benefits international capital holders, replicating a trite colonial relationship between peoples and governments, producers and consumers, polluted and clean neighborhoods. The pollution of water sources, the drying of rivers, or toxic waste on Indigenous (and Black) bodies are scenarios that repeat much too often (Flint still has no clean water!). The postcolonial state's racism is systemic everywhere, with a global accumulation of capital built on cultural and territorial dispossession of Indigenous (and Black) nations. Scholars, too, engage in a "refusal to articulate the immanence of intellectual life and the concrete geohistorical determination of ideas" (Lionnet and Shih 2011, p. 15). So that a "concrete thinking of difference" is pivotal to intercultural relations that counter forced acculturation and epistemicide (Lionnet and Shih 2011, p. 27).

In *The Land is Our History* (2016), Miranda Johnson tells the story of a 1963 Indigenous petition sent to the Australian parliament in English *and* in Yolngu, the language of Indigenous nations in Northern Australia, demanding to protect their land from mining companies, land that is "vital to their livelihood" (2016, p. 15). This story finds echoes everywhere as consequence of settler law in a British postcolonial commonwealth. Indigenous peoples in Australia, Canada, and the United States began to organize in the framework of civil rights and antiapartheid models, imagining a future that was not assimilationist. Like *Sumak Kawsay* and *Pachamama* in the Andes, Johnson's Australian example shows a semantic dissonance between the place described as land when used by the elders and the government's use of the word. Yolngu's land rights include revival and self-determination. Land rights still provides the space to set struggles using a common language of transnational political modernity, collective rights specific to Indigenous citizenry (Johnson, 2016, p. 30).

In *The Climate of History in a Planetary Age* (2021), Dipesh Chakrabarty takes up Carl Schmitt's idea of today's globalized capitalism as a spaceship that is the technosphere we live in. Global corporations trawl the ocean, mining companies move billions of tons of dirt. Humans reshape the planet. Mining is one industry where the human and geological histories connect to set in motion not only Indigenous genocide and ecocide but a planetary mass extinction (Chakrabarty, 2021, p. 7). We live on a planet where hills and glaciers go "missing" due to a global capitalist extractivism that creates a "planetary environmental crisis" whereby the political inevitably extends to Nature (Chakrabarty, 2021, p.13). The "ravages of extractive capitalism" unequally affect racialized communities in an uneven development built on colonial models of oppression (Chakrabarty, p. 15). Chakrabarty's urge to extend political modernity to Nature cannot be satisfied without a serious examination of our social systems and their structural forms of inequality that refuses to recognize the political modernity of racialized nations.

These past decades reshaped the intercultural relations that inform the Canadian treatment of Indigenous peoples, and of Indigenous nations in Ecuador, now subject to extractive models. According to an Organization of American States's report, in 2019 alone, Ecuador exported $76.7 million of ore to China, though China reported $339 million imports from Ecuador, so the report concludes this gap is made of illegal mining that far exceeds the legal industry (OAS, 2021, p. 34). In Colombia and Peru, illegal mining and drug export networks have produced "cross-border flows" into Ecuador which, in 2019, exported four times more ore gold to China than Colombia and Peru combined. COVID exacerbated an already severe problem for racialized communities defending water and nature from an extractivist development. An uneven development affects the Amazon and Esmeraldas in specific ways. In the last decade, mining has grown exponentially in Ecuador and the current president, Guillermo Lasso, has given at least 170 new mining contracts to Canadian corporations.

The Living Forest

Yaku's continental encounters continued to take place as regional organizations solidified in a concerted effort to resist extractivism transnationally. At the 2023 Third Continental Encounter, I translated for the Amazonian delegations. Josi Otaviano Guilherme, a Tikuna water defender from the triple border between Peru, Colombia, and Brazil, denounced in Portuguese her people's invisibility, underrepresented by half in the official census as opposed to their own communitarian counts. She saw her activism as inevitable: "the forest is dying" [*a foresta está morrendo*] (Somos Agua, 2023). The 2015 uprising of Brazilian Amazonians she led was for the defense of water, the forest, the territory, mother earth, the seeds, and of ancestral knowledge. Meanwhile, women from the Colombian Putumayo defended their "territory for life" [*territorio de la vida*], a similar concept to *Sumak Kawsay*, as the site for the reproduction of life and life technologies.

Postcolonial governments disregard the ILO Convention 169, the 1989 document that established international standards for the guarantee of fundamental human rights for indigenous peoples. Like in Azuay, the Brazilian and Colombian leaders denounced a denial of previous consent (*consulta previa informada*). Without it, no Indigenous nation can have autonomy or legal guarantees. Putumayo women conclude, there is no effective participation in the processes that generate such immense impact in Amazonian Forest territories. The Indigenous Amazonian nations share a language and in the Colombian side of the border, they too used *Sumak Kawsay* as a legal instrument in their petitions. From their house of water, *Yakuwasi*, Putumayo women demand that the government take immediate action to mitigate the water crisis and call for an international observatory that can force restorative justice on damaged water ecosystems (Somos Agua, 2023). In 2016, Colombia's Supreme Court recognized Nature as a rights-bearing subject in a country without any constitutional protection. In fact, the rights of nature and for life are now recognized in many countries around the world like Panama, Bolivia, New Zealand or India, thanks to international law.

The demands for the rights of nature are clearly influenced by the Ecuadorian model that has begun to be tested internationally in South America and the world. The novel 2008 Ecuadorian Constitution is often attributed to the white, handsome, charismatic former president of Ecuador, Rafael Correa, in his attempt to take Ecuador out of its "long night of neoliberalism" [*larga noche neoliberal*] (Becker, 2015, p. 3), but it was in fact the result of decades of conceptual work of Indigenous nations in constant struggle against the state. Marc Becker explains how the *Pachakutik* movement took a protagonist role in electoral politics in the 1990s and was an important pillar to advance concrete solutions to long standing Indigenous demands such as the defense of land, language, and autonomy. Indigenous nations adapted each decade to the obsolete shapes of postcolonial politics to maintain their demands on the negotiating table with the state (Becker, 2015).

The Rights of Rivers

According to Pablo Piedra, defending lawyer for Fierro Urcu, in the southern province of Loja, mining deregulation began with Correa in 2008 with a model of "extractivist extremism" instead of dialogue, and the rule of violence in the absence of virtue (Somos Agua, 2023). In Tundayme, then-president Lenin Moreno also gave mining contracts without previous informed consent. In 2020, he argued that it was easier to ask for forgiveness than permission, words that triggered deep painful memories in Zamora, where the extermination of Indigenous peoples for early colonial mining brought slavery into the region. Mining Fierro Urcu would affect the Podocarpus National Park that crosses provincial borders between Loja and Zamora. Nature's constitutional rights were weakened.

Fierro Urcu now has different mining concessions which violate constitutional rights of peoples, nature, and water, among them Guayacan Gold, Cornerstone, and SolgGold. The 2020 petition was to protect the river system, the largest in the southern Andes, bringing water to four counties and dozens of towns where the Indigenous Saraguro live. On May 22, 2022, a group self-identified as the Guardians of Fiero Urcu denounced the paramilitary presence in Loja's *Defensoría del Pueblo*; two days later, Saraguro women blocked the entrance to the tundra at their golden hillside with a clear demand, *"nosotras queremos nuestro cerro sagrado intacto para vivir"* [we want our sacred mountain left intact for life] (Montaño, 2022). They further spoke of an invasion of their ancestral territory by police and paramilitary forces, again tapping into a long memory of an army headed by dogs, ravaging Indigenous lands and people for resource extraction. The support of scientific reports for its unique ecosystems and biodiversity has been key to protecting Fierro Urcu, where one hundred people have already been criminalized for defending the planet with both Indigenous and environmental activism.

In the Northern Andes, the mining scenario repeats. Agustín Grijalva explains how Intag has 117 different orchid species in a forest where over 70 percent of its acreage has been allotted for open-pit mining (Somos Agua, 2023). The town of Cotacachi along with social organizations prepared a legal action to protect Intag's forest and rivers, but the local judge rejected the motion. The appeal to the provincial court was supported by sworn scientific reports in defense of ecosystems, *amicus curiae*, and accompanied by massive protests. The case went all the way to the Supreme Court where the motion passed, but judges did not uphold the rights of nature, particularly the rights of four rivers affected by the project *Los Cedros*. Ecofeminist Esperanza Martínez goes further and speaks of the bioterritory, which comprised much more than the flat surface featured in maps (Somos Agua, 2023). She speaks of nature reacting in a vertical way to the crimes against her, against pollution of the water in rivers, the rain, the tundras, clouds, and flying rivers—water "indivisible from the mountain to the sea." Its pollution affects life in every dimension. She concludes that environmental justice has failed, so we must legislate the rights of nature, not conceived as the environment, but as subject of law.

For the Ataco River, the biocultural elements of the river are as important as the rights of nature, since there is a close and interactive relation between peoples, spiritual meanings, and way of life. The expropriation of Indigenous resources results in natural catastrophes, like the dam that polluted the Coca River. Megaprojects such as highways, dams, and mines cause damage that demands reparative justice, though prevention, an antimining law, and a love for the rivers big enough to rename them in an attempt at restoration would be ideal. Why call the Colombian river, trash—*Río Basura*, or the one in the Amazon, bad—*Río Malo*, or the one in Esmeraldas dirty—*Sucio*? Rivers also have the right to a memory that is not colonial and extractivist (Somos Agua, 2023). Ramiro Ávila explains how court rulings can have a significant impact on river restoration. He describes how the Monjas river in the North of Quito was "vomiting mercury" and, like many others, was full of excrement that Quito's municipality did not care to treat. Judge Ávila made history with his ruling granting legal personhood to the Monjas river, an unprecedented move that he then replicated in his decision over the pollution of the Aquito river in Tsáchila territory. In 2022, Ávila found Quito's townhall guilty of infringing upon its citizens' right to "live in a healthy environment . . . to a safe habitat, to water, to a sustainable development . . . the rights of nature, and to a cultural heritage" (Ávila, 2022, p. 1).

Special UN Rapporteur, Francisco Calí Tzay reflects on the extractivism of Indigenous resources, given that these nations are not only protecting nature, but bioengineering life. It is no coincidence that 70 percent of all the planet's biodiversity is circumscribed within Indigenous lands. In his 2022 report, he acknowledged that women in particular play a significant role in the protection of ancestral knowledge, which includes agricultural techniques, medicine, the arts, endangered languages, textiles, and spiritual elements that combined can mitigate climate change with resilience (Somos Agua, 2023). But Indigenous nations have no financial resources to revitalize Indigenous epistemes or protect the forests, especially when faced with the "strategic sector" discourse and its human and environmental impact.

What water and nature protectors agree on is that governments do not respect informed, previous consent, political frameworks do not match indigenous visions, and the violence against defenders shows a lack of respect for collective rights and a legal means for self-defense. The misinformation campaigns and the rewriting of history make things worse. For example, on February 26, 2023, Cofan leader Eduardo Mendúa, CONAIE's (*Confederation of Indigenous Nationalities of Ecuador*) Director of International Relations, was gunned down in his home at Dureno for openly resisting the strategic sector discourse to take A'i Cofan lands for oil drilling, which he considered theft (Etchart, 2023). His wife, Fabiola Ortiz led the feminist 8M march where she said "they tried to silence my husband's voice, but they didn't know that they lit many more. Long live our resistance!" [*Quisieron apagar la voz de mi esposo, pero no se dieron cuenta que encendieron muchas voces más. ¡Qué viva la lucha!*] (Etchart, 2023). On June 1, 2023, Josi Tikuna's family compound was attacked, causing the death of a family member. Her mother is still missing but feared to be dead. In the Kimsakocha case, I had to prepare an *amicus* to explain that the Kañari people are in fact Indigenous and live in the areas affected by mining. The government and mining companies had argued through paid expert reports that the people living in this tundra were not Indigenous, despite having an Indigenous language, name, and dress. Naturally, the historical record proved them wrong. If the Kañari do not exist, the most fundamental facts about Ecuadorian history would be denied, like the Kañari helping the Spanish beat the Inca Empire in the sixteenth century. This case is still pending. Yaku and many other Indigenous defenders in Australia, Canada, New Zealand, Latin America, and even in Northern Europe must use international courts to force postcolonial governments to recognize their rights and stop Indigenous denial. Using international courts is an important strategy, so the fifty-seven UN rapporteurs on human rights help provide tools to build transnational agendas for the defense of the rights of nature and Indigenous political modernity.

Surviving Dystopia through Indigenous Alliances

When former president Velasco Ibarra invited American oil companies to destroy the Amazon, his model included missionary partitions of the forest, the most notorious case being the Summer Institute of Linguistics, where five missionaries were killed by the Waorani in 1958 (Carcelén Estrada, 2010). Marc Becker explains how in the Southern Amazon and in the context of the Agrarian Reform, the Shuar allied with Salecian missionaries influenced by the theory of liberation, while in the Northern Amazon the Naporuna organization, FEPOCAN (*Federación de Pastaza de Organizaciones Campesinas*), worked with Josefine missionaries since 1969 (Becker, 2015). In 1960, the FENOC (*Federación Nacional de Organizaciones Campesinas*), a catch-all organization, was set up by the Catholic church to try to minimize the work of the communist federation, the FEI, founded in 1944, that had brought the infamous Tránsito Amaguaña to give a speech in Moscow and Velasco Ibarra to speak at the national congress in Kichwa. Missionaries supplanted communism as intermediator between Indigenous and national politics. In 1996, the FENOC became the FENOCIN (*Confederación Nacional de Organizaciones Campesinas Indígenas y Negras*) and included intercultural demands in their agenda (Becker, 2015, p. 22), which today includes the Montuvio ethnic group as well as Indigenous and Black organizations.

During the Ecuador's model of oil development through the missionary colonization of the Amazon, the term federation started to be used to designate Indigenous movements. In the 1970s these federations brought together working committees, operating separately from the church, such as: the FOIN (*Federación de Organizaciones Indígenas*), founded in 1973 to rethink land relations to property and to advocate for an ancestral worldview of language and territory; or the more "militant" ECUARUNARI (*Ecuador runacunapac riccharimui*), founded in 1972 from the *campesino* networks of Chimborazo, Cañar, and Azuay in the Southern Andes (Becker, 2015, p. 10). In 1978, in the Amazonian city of Puyo, the FECIP (*Federación de Pueblos Indígenas de Pastaza*) was established to advocate for the environmental rights of the forest (Becker, 2015, p. 9). By 1980, Amazonians and Andean Indigenous ethnicities began to meet to talk about the environment and clearly articulate their political agenda around cultural, linguistic, and territorial rights for self-determination. To do this, they founded the CONFENAIE (*Confederación de Nacionalidades Indígenas del Ecuador*) in Sucúa where they were already talking about a plurinational state as was conceived in the Amazon, an idea propagated through radio and bilingual education (Becker, 2015, p. 11). When in 1986 the CONFENAIE (*Confederación de Nacionalidades Indígenas del Ecuador*) was formed, its first president, an Amazonian Shuar, Miguel Tankamash, pushed for an Amazonian model of plural citizenship and a plurinational state that defends and valorizes Indigenous rights, cultures, and territories (Becker, 2015, p. 11). The CONFENAIE also included an international demand: to reestablish diplomatic links with the revolutionary Sandinista government of Nicaragua (Becker, 2015, p. 14).

An Amazonian Plurinational and Andean Intercultural Citizenship

For the Amazonian Shuar, plurinational means the defense of their ancestral territory against the state, church, and corporations, but for FENOCIN, the demand was also for interculturality not as a space for discussing difference, but for achieving unity (Becker, 2015, p. 16). These competing notions of a plurinational and an intercultural society were set aside in a strategic Andean-Amazonian alliance that was at its core anticolonial, anticapitalistic, and antiimperialist (Becker, 2015, p. 20). When a national Indigenous uprising paralyzed the country for the first time in 1990, the Shuar had worked for decades on their version of a new world that could make room for difference, not just between mestizos and Indigenous populations, but also among different Indigenous ethnicities, who already called themselves nationalities (*nacionalidades*), resignifying community away from Marxist meanings and into an antisystemic model based on a plurinational recognition of culture, the economy, language, a politics of autonomy, self-determination, and the defense of the territory.

The birth of participatory democracy, then, occurs in the Amazon, in Tena, Puyo, Sucúa where the FENOCIN, FEI, CONFENAIE, and CONAIE took shape, defined goals, and organized strategies with plenty of room for difference that respected the Andes and the Amazon as distinct political entities with competing meanings of democracy and of the territory hidden behind the umbrella concept of interculturality. The reconciliation of Indigenous difference through interculturality, lead to the political model of a plurinational state combining an older Andean model of Marxist organizing with the intersectional and multidimensional Amazonian model that included the entirety of differences and a new democracy for all. Correa distorted this democratic model by resignifying its organizational meanings, including *Sumak Kawsay*, interculturality, and sustainable development.

The plurinational state for the Shuar was a strategic alliance among Indigenous peoples at a national level, meant to turn oil extraction on its head. And it worked. Throughout the 1990s, significant gains like a constitutional recognition of ancestral lands satisfied some Indigenous demands, but they also managed to capture the mestizo imagination when urban protesters began using their anti-neoliberal discourse during the privatization years of Sixto Durán Ballén (1992–1996) (Becker, 2015, p. xv). Only then did the Indigenous movement go truly national through the interculturality for strategic alliances, leading to the formation of the political wing of the CONAIE, *Pachakutik*, which literally translates as the upending of time-space, or the new beginning of an Indigenous political time. *Pachakutik* starred in congress and was a protagonist in the overthrows of presidents Abdalá Bucarám in 1997, Jamil Mahuad in 2000, and Lucio Gutierrez in 2005 (Becker, 2015, p. 20). Isolated in his ivory tower, then Professor Rafael Correa had nothing to do with this process of change.

Becker demonstrates that Indigenous political action is necessarily tied to this "*campo minado de las identidades étnicas y polítcas*," the mined field of ethnic and political identities (Becker, 2015, p. 17). Interculturality´s double meaning perhaps represents the competing meanings within a very complex and diverse indigenous world of worlds. For the Amazonian model of a new society, interculturality opens up a space for difference, a difference between many Indigenous nations and their respective postcolonial experience as either Kichwas (Andean) or Shuars (Amazonian), the main nations behind the formation of a national Indigenous movement that was intersectional (class, gender, and ethnicity), plurinational (respecting linguistic and cultural differences), and intercultural (in constant dialogue with the allied citizens, the state, the world) (Becker, 2015, p. 15). These competing meanings do not only reflect separate agendas, but distinct overlapping dimensions, *"la doble dimensión de nuestra lucha"* (Becker, 2015, p. 10). As nations and as Andean and Amazonian, newly conquered or old servants to a colonial state, with local histories and global demands, these nations came together to strategically align against inequality and overcame internal divisions.

Rafael Correa (2007-2017) renewed extractivism in the new millennium with a leftist discourse that Thea Riofrancos calls "radical resource nationalism," which she identifies as dichotomous to Indigenous anti-extractivism (Riofrancos, 2021, p. 3). This is a false dichotomy that fails to understand the multidimensionality of Indigenous resistance. While it is true that Correa expanded oil exploitation to execute cash transfers and reduce poverty in a traditional Latin American Marxist sense, the "radicalized" Indigenous resistance was fighting extractivist corporations regardless of flag, including Chinese companies, which together took three quarters of Ecuador's primary resource production from 2000 to 2010, when oil represented 50 percent of the government income and one third of its expenditure (Riofrancos, 2021, p. 7). Correa sold oil to China in exchange for loans: 17 trillion to Petroecuador alone in 2017 (Riofrancos, 2021, p. 9). He also gave new mining contracts for copper and gold exploitation in the Southern Amazon, made extractivism a national priority, and declared extractive industries to be strategic sectors.

These contracts included Mirador in Zamora Chinchipe, San Carlos-en-Panantza in Morona Santiago, and thirteen new pits for oil extraction. Correa then attempted a legal reform to overhaul constitutional protections, a trend shared with Argentina, Bolivia, and Venezuela (Riofrancos, 2021, p. 9). The New Left, or the Pink Tide or the Twenty-First Century Socialism, whatever label is preferred, did nothing to restructure colonial power relations, but instead, in Ecuador, the Mining Laws (2009) and Decree 870 (2011) imposed the construction of infrastructures in Indigenous communities, infrastructures built with Chinese money paid for with "anticipated royalties," which amounted to 100 million dollars sent to China with the Mirador project (Riofrancos, 2021, p. 14). Riofrancos concludes that the division of the left (labor, peasant, environmental, and

Indigenous movements) resulted in an incapacity to unify all fronts under a single anti-extractivist movement (Riofrancos, 2021, p. 15). Meanwhile, extractivism remains the law of the land for the ruling elites.

Black Women in Resistance

Postcolonial spaces and peoples get redefined as imperial borders advance. Esmeraldas, as a "savage jungle," takes on a meaning of place that widely contradicts communitarian conceptions of rivers and mangroves as a web with their own connotations and constellations for Black re-existence in the Chocó, a region that runs through the Ecuadorian and Colombian Pacific. About half of Esmeraldas residents currently self-identify as Afro descendant. Today, racialized women demand a Black memory that goes deep into a feminine experience of the Pacific to tell a story with its own codes (Carcelén-Estrada, 2022). Esmeraldas's transnational resistance through orality and an aesthetics of resistance to defend life and their lands finds echoes in the Black experience of women in Brazil, the Caribbean, and the United States, particularly in the context of the Black Lives Matter movement. The extractivist racist footprint and the "spaces of sacrifice" (Moreano, 2019, p. 59) is transnational, and so is its resistance.

Black women's libertarian thinking and their demand for justice have always been at the vanguard of liberation processes. *Palenquera* female memory is rooted in libertarian practices already in place that are strengthened through a sustained dialogue with the Caribbean, Brazil, and North America, all immersed in a Black transnational feminism. For Angela Davis, abolition is a continuing process because so do lynchings, murders, disappearances, police violence, and the militarization of daily life (Davis 2004 [1981], p. 95). She has fostered the creation of spaces for the development of Black feminist theory through meetings and making her work available through translation. More recently on September 7, 2021, the People's Forum organized a dialogue via Zoom between Davis and Francia Márquez, vice president of Colombia. They discussed the African epistemic praxis of Ubuntu—I am because we are, and no one can be when in isolation—in a post-Covid world that isolated all of us (The People's Forum, 2021). The translatability of interwoven Black concepts like *cimarronaje*, Ubuntu, Afro-descendant, raizal (unruly rootedness), or creolization, opens spaces of encounter for transnational Black feminist activists. Like Angela Davis, there are innumerable black women transforming the social space through their political thinking and action in translation.

Black women's transnational struggles and the paths that Davis has taken, from the Chocó to Bahía, are only a sample of a liberation network that underlines dispersion and archipelagos of Black transnational feminist activisms in Abya Yala. Decolonial Black feminist methodologies craft a story of rejoicing, like Francia Márquez's *vivir sabroso*, the joyful living many activists speak of when translating Ubuntu into the *bodies-territories* of the continent. This is not a translation of imposed concepts, but a space for the healing of *bodies-in-encounter* with methodologies in-place (Carcelén-Estrada, 2022). Ubuntu orally translates into a political praxis in-place, without translation outside of its own space of practice, *casa afuera*, in spaces of encounter that a mestizo audience constantly misunderstands due to their structurally racist and violent gaze. The untranslatability to a white audience seems irrelevant when it concerns a resignification from within, a closed intralingual translation, *casa adentro*.

Necropolitics in the Chocó

National development has caused much violence in the Colombian Chocó and includes dispossession, paramilitary attacks, dismemberments, water graves, humiliation, and disappearances, among others (Lozano, 2019, p. 32). Feminicide as a public manifestation of violence that arises from national development is just one example of how the state or paramilitary armies display control over the bodies of Chocoan women, where the fight over the control of the Chocó takes place. Displacements in Cali, Medellín, Bogotá (as in Esmeraldas, Guayaquil, and Quito) are territorial extensions of the Palenque, carving freedom spaces for entire migrating communities and their cultures. Border violence follows them into the slums of cities, into jails, schools, everywhere where they are not considered as humans defined beyond race. The structural violence of national development adds to gender-based violence on the ground leaving individual and communitarian wounds, an accumulation of violence that puts "everyone against each other" (Lozano, 2016, p. 33). In this context, women's struggles shape concepts for resistance and re-existence and a dignified life. Even when her life is in danger, since I met her in 2008, Francia Márquez continues to lead struggles that advance everyone's fundamental rights. Black solidarity networks turned her into a public international figure that could shield her personhood from targeted state and paramilitary violence. Her defense of La Toma from transnational mining companies led her to become Colombia's first Black vice president in the first leftist government in seventy years of violence.

Black transnational feminisms weave a multiplicity of resistance in translation in its three colonial languages, English, Spanish, and Portuguese. Black activists certainly act as "political translators" disrupting the dominant discourse and expanding democracy (Doerr, 2019, p. 189). To do this, they use international law and the United Nations to force governments to take decolonizing steps. Thanks to Black women, international institutions guarantee Afro-descendants rights, institutions like the Organization of American States (OAS), the Office of Rapporteurship on the Rights of Persons of African Descent and against Racial Discrimination (OAS, 2005), the Inter-American Development Bank's division on Gender and Diversity (Banco Interamericano de Desarrollo, 2007), and the Committee on the Elimination of Racial Discrimination (CERD) under the Office of the High Commissioner of Human Rights (OHCHR), a UN program since 1965. Women capture the ontology of their pluricultural identities in the religious festivities they organize, their agricultural production, and the tending of medicinal plants. They also recreate their ecosystems and natural economy, both threatened by an extractivist development behind shrimp farms, large-scale mining, and coca plantations. Their racialized territory remains trapped in the colonial chronotope of slavery, even though Esmeraldas was birthplace of national liberation and the setting for Ecuador's longest war, *La Guerra de Concha* (1913–1916). Despite this history, school manuals emphasize the railroad, cocoa and coffee production, and banana, shrimp, and palm oil farms, all industries complicit with colonial dispossession and essential for an extractivist export economy.

Afrodiasporic feminism is "a complex, contradictory, broad, and heterogeneous perspective for the political thinking, action, and life that emerge from realities in different historical moments and geographic spaces where Black women are protagonists" (Vergara-Figueroa and Arboleda, 2014, p.113). Black women's lived experiences are political theory. Black women recreate life and death in the Pacific despite the neoliberal impoverishment of necropolitical state policies: hospitals shutdown, lack of doctors, no running water, schools, or food sovereignty. Extracting knowledge, minerals, trees, mangroves often results in forced displacement caused by pollution, the destruction of habitats for traditional fishing economies, and state and paramilitary violence that often accompanies a borderland ready for extraction. Mara Viveros argues that, in the Route of the Pacific, the paths of re-existence converge in translation in the concept of *body-territory* (Mara Viveros, 2002, p. 127), defined as a link to the planet, rooted in territorial relationalities of a very specific experience (Mara Viveros, 2017). The concept *body-territory* is a category that emerged from the work of Indigenous and Black women across Abya-Yala, with the objective of generating knowledge in-place, resignifying the political through the organization and reproducing communal life, mutually protecting themselves from state violence (Tzul-Tzul, 2015, pp. 92, 98). Astrid Ulloa applies these decolonial categories to her work with the Wayuú nation in Colombia, who speak of a coexistence with space and with species as a social web in the aftermath of mining devastation (Ulloa, 2014, p. 6–15).

Structural racism must be made visible. Communities already gather data on their own and disseminate information on the inner workings of structural, institutional, state, and interpersonal racism. The UN World Conference against Racism, Racial Discrimination, Xenophobia, and Related Intolerance that took place in Durban, South Africa in 2001 normalized the Afro-descendant label in South America, where movements were most successful in implementing Durban's agreements, which included an Afro descendant identity. Yet, despite the significant strides in international and constitutional law, racial inequalities still widen in the South American postcolonies. As in the many spaces the Afrodiaspora inhabits, Ecuador punitive colonial geography continues to criminalize the Palenque while a racist educational system is complicit with Afro-descendant epistemicide. To fight against an extractivist development and restore Esmeraldas, women similarly recompose communities and ancestral memory with codes and symbols that are untranslatable to a violent capitalist and genocidal state. Esmeraldas diasporic memory travels with the displaced to big cities, while women recompose the Pacific's image and meaning, mapping their defense against an extractivist development that produces an ever-wandering, untranslated diaspora.

For activists in the Chocó, misunderstandings are a given. An attempt to dialogue with whiteness seems futile, categorically impossible. Translation leads to an epistemicide built on colonial power structures, where the Global North marginalizes the Global South privileging a disembodied knowledge production over a mistrusted "theory that takes up embodied experience" (Price, 2023). Global ideas travel and adapt as activist translators make translation visible (Muller, 2007). Global South feminist concepts, methodologies, and practices increasingly creep into the North Atlantic in translation into English, often domesticating and depoliticizing women's political action in Peru, China, or Ethiopia, particularly in contexts of development (Levitt and Merry, 2009; Ostebo, 2015). At best, "translation of global norms into domestic practice remains principally incomplete" (Zwingel, 2017, p. 10). To prevent dispossession, or becoming an ornament to development discourses, untranslatability seems an appropriate response. This South-to-South conversation is untranslatable to a complacent Global North. So, where does translation take place? Among feminist activists in intercultural spaces of Black memory speaking creole languages with messages that expand human, political, and social rights for all.

Untranslatability seems adequate to resist epistemicide. Conversations occur orally, creolized in places of knowledge production that the Global North has a hard time imagining. Structurally and intrinsically racist and colonial, Ecuadorian mestizos, too, have a hard time imagining border communities as sites of knowledge production, so why try to explain? The province of Esmeraldas, like the rest of Ecuador, is coming out of an emergency health crisis that has seriously weakened the economy and social stability of its citizens, especially among those who are unemployed and live day by day. But when the health conditions seemed to improve for the rest, Esmeraldas came out to be terrorized by criminal groups with total control to carry out illicit activities, such as drug trafficking, extorsion, hitman killings, and so on.

According to the Secretary of State and the National Police, in 2021, murders in Esmeraldas increased to 151. In 2022, this number more than tripled to 520 violent deaths, and there were already 144 deaths in the first trimester of 2023. Nationally, in the first month of 2022 there were 416 reported violent deaths. By the end of March 2022, there were 815 reported violent deaths. In the same period in 2023, 4603 victims fell to this unleashed violence, an increase of 66.4 percent (La Hora, 2023, p. 8). Eighty-five percent of all deaths take place in just fifteen districts, all Black. The county of Esmeraldas reports the highest murder rate of the seven counties in this province. Most of the deceased had criminal records, so Javier Buitrón Flores, the commander of the Esmeraldas district No. 8, calls this violence a "quarrel between gangs" and Juan Zapata, Minister of Interior, says that since this is a "war between gangs" there is nothing the government can do to stop the killings (La Hora, 2023). But how then to explain the murder at the local university Luis Vargas Torres of its dean of Economics, Segundo Castillo Cabezas? Esmeraldas's necropolitics and the government's utter disregard for Black lives is evident in epistemicide, ecocide, ethnocide, and now blatant genocide. If 2022 seemed to point to a violence out of control with an increase of 82.5 percent since 2021 (González, 2023), the current murder rate made such numbers look small. If this trend continues, there will be 6300 violent deaths by the end of the year, making Esmeraldas the most violent place on earth. #SOSEsmeraldas.

Dystopian Realities

Genocide, ecocide, and epistemicide are closely related. In Esmeraldas, decades of mining, shrimp farming, deforestation, and the destruction of the mangrove have made the transmission of African knowledge difficult, especially when impeded by migration. Genocide makes this migration worse, and ecocide makes this land especially vulnerable to natural catastrophes. This is what happened on June 5, 2023, when seven rivers, covered by mangroves in the past and now barren, overflowed, causing the biggest flood in Esmeraldas's history, with 15,000 people displaced. These rivers are the Súa, Tonchigüe, Cube, Viche, Teaone, Blanco, and Metambal. Meanwhile the fascist government responds with the same tools as it does to violence, *estado de excepción*, state of emergency to spend millions of dollars without oversight, and with no real benefits for the local communities. President Guillermo Lasso has kept the province in a permanent state of emergency, with no solutions and in complicity with drug lords who are allowed to operate freely. The narcostate also favors mining, which continues to increase in Esmeraldas and in the Amazon, like in Punino where there was a 578 percent increase in mining in just one year (2021–2022) (#NAPORESISTE). After he dissolved congress, Lasso was given six months to rule by decree, and it is expected that these months will see an expansion of mining with disregard for the Constitution that was put on hold until the next round of elections. The fascist state sponsors deforestation and mining that leads to climate catastrophe, leaving racialized communities with a pessimistic outlook for the future. Transnational resistance is necessary for the defense of life and water. This is what the COICA (Confederation of Indigenous Organizations of the Amazon Basin) advocates, concerned with mining in Peru and Ecuador, and now engaged in deep conversations with the Colombian and Brazilian Amazon.

References

Ávila, R. (2022). Corte Constitucional del Ecuador, 2022. *Sentencia No. 2167-21-EP/22.* Available at: http://esacc.corteconstitucional.gob.ec/storage/api/v1/10_DWL_FL/e2NhcnBldGE6J3RyYW1pdGUnLCB1dWlkOic5OWVmN2EyZC1kM2I5LTQwOWQtOWY4ZS1jMDc3YzYxYWQ2ZGMucGRmJ30=

Becker, M. (2015). *¡Pachakutik! Movimientos indígenas, proyectos políticos y disputas electorales en el Ecuador.* Quito: Abya Yala.

Carcelen-Estrada, A. (2022). Oral Histories in the Black Pacific: Women, Memory, and the Defense of the Territory. *Radical History Review* 144, 22(3) pp.77-105.

Carcelen-Estrada, A. (2010). Covert and Overt Ideologies in the Translation of the Wycliffe Bible into Huao Terero. In Tymoczko M. (Ed), *Translation, Resistance, Activism*, pp. 65-88. Amherst: University of Massachusetts Press.

Chakrabarty, D. (2021). *The Climate of History in a Planetary Age.* Chicago: The University of Chicago Press.

Davis, A. (1981; reprint: 2004). *Women, Race, and Class.* Madrid: Akal.

Doerr, N. (2019). Activists as political translators? Structural Inequality, Gender, and Positionality Misunderstandings in Refugee Solidarity Coalitions. In Irvine J., Lang S. and Montoya C. (Eds.), *Gendered Mobilizations and Intersectional Challenges: Contemporary Social Movements in Europe and North America*, pp.189-207. London: Rowman & Littlefield Publishers.

Etchart, L. (2023). The Assassination of Eduardo Mandúa. *Latin American Bureau*, March 15. Available at: https://lab.org.uk/the-assassination-of-eduardo-mendua/

González, M. A. (2023). Ecuador lidera el crecimiento de violencia criminal en Latinoamérica. *Primicias*, January 16.

Johnson, M. (2016). *The Land is our History: Indigeneity, Law, and the Settler State.* New York: Oxford University Press.

Esmeraldas. (2023). *La Hora*, March 29, 2023. Available at: https://www.lahora.com.ec/edicion-digital-pdf/edicion-del-dia-esmeraldas/esmeraldas-29-de-marzo-2023/

Levitt, P. and Sally M. (2009). Vernacularization on the Ground: local uses of global women's rights in Peru, China, India, and the United States. *Global Networks*, 9(4), pp.441-461.

Lionnet, F. and Shu-Mei S. (Eds), (2011). *The Creolization of Theory*. Chapel Hill: Duke University Press.

Lozano, B. R. (2016). Feminismo negro-afrocolombiano: ancestral, insurgente y cimarrón. Un feminismo en-lugar. *Intersticios de la política y la cultura Intervenciones Latinoamericanas*, 5(9), pp. 23-48.

Montaño, D. (2022). El caso Fierro Urco, explicado. *GK*, June 2. Available at: https://gk.city/2022/06/02/caso-fierro-urco-explicado-estrella-hidrica-sur-mineria/

Moreano, M. (2019). Geografía marxista y materialismo histórico geográfico: Más allá de la acumulación por desposesión. *Geografía Crítica para detener el despojo de los territorios*, ed. Colectivo de Geografía Crítica. Quito: Editorial Abya-Yala.

Muller, M. (2007). What's in a Word? Problematizing Translation Between Languages. *Area*, 39(2), pp. 206-213.

Organization of American States. (2021). *On the Trail of Illicit Gold Proceeds: Strengthening the Fight against Illegal Mining Finances*. Washington DC: Department against Translational Organized Crime.

Organization of American States. (2011). *Constitución de la República del Ecuador 2008*. Legal documents, July 13, 2011. Available at: https://www.oas.org/juridico/pdfs/mesicic4_ecu_const.pdf

The Inter-American Commission on Human Rights, (2005). *Report of The Office of Rapporteurship on the Rights of Indigenous People*. Available at: http://www.cidh.org/annualrep/2005eng/chap.2a.htm#discrimination

Ostebo, M. (2015). Translations of Gender Equality among Rural Arsi Oromo in Ethiopia. *Development and Change*, 46(3), pp.442-463.

Pérez Guartambel, Y (2012). *Agua u Oro. Kimsakocha, la resistencia por el agua*. Cuenca: Universidad de Cuenca.

Price, J. (2023). *Translation and Epistemicide: Racialization of Language in the Americas*. Tucson: University of Arizona Press.

Riofrancos, T. (2021). *Resource Radicals*. Chapel Hill: Duke University Press.

Somos Agua. (2023). III Continental Encounter for Water and Climate Crisis. Trans. Antonia Carcelen-Estrada. Personal translation notes. Available at: https://www.defensoresdelagua.com/encuentro/.

Simon, S. (2018). Introduction, Special issue of *Translation, a transdisciplinary journal* 7(1), pp.11-15.

The People's Forum. (2021). I am because we are. *The Center for Place, Culture and Politics.* September 7. Available at: https://peoplesforum.org/events/i-am-because-we-are-a-conversation-between-francia-marquez-mina-and-angela-davis/

Tzul-Tzul, G. (2015). Mujeres indígenas: Historias de la reproducción de la vida en Guatemala. Una reflexión a partir de la visita de Silvia Federici. *Bajo el Volcán,* 15(22), pp. 91-99.

Ulloa, A. (2014). The Rights of the Wayúu people and Water in the Context of mining in La Guajira, Colombia: demands of relational water justice. *Human Geography,* 13(1), pp.6-15.

Vergara-Figueroa, A. and Arboleda, K. (2014). Feminismo afrodiaspórico. Una agenda emergente del feminismo negro en Colombia. Universitas Humanística, 78(78), pp:109-134. Available at: http://dx.doi.org/10.11144/Javeriana.UH78.fafn

Vivero, M. (2017). Intersecciones, periferias y heterotopías en las cartografías de la sexualidad. *Sexualidad, Salud y Sociedad,* (27), pp. 220-241.

Vivero, M. (2002). *De quebradores y cumplidores. Sobre hombres, masculinidades y relaciones de género en Colombia.* Bogotá: CES.

Zambrano, M. (2002). De quebradores y cumplidores. Sobre hombres, masculinidades y relaciones de género en Colombia. *Revista Colombiana de Antropologia,* 38, pp.329-332.

Zwingel, S. (2017). Women's rights norms as content-in-motion and incomplete practices. *Third World Thematics,A TWQ Journal,* 2(5), pp: 1-16. Available at: https://doi.org/10.1080/23802014.2017.1365625

Part 3

Taking Back Indigenous Lands, Water and Epistemes

Chapter 9

Will Wilson: Connecting the Dots of Uranium Extraction on the Navajo Nation

Kaila T. Schedeen

Portions of this essay were originally published in the exhibition brochure for Will Wilson: AIR/Survey, *held at the Visual Arts Center at The University of Texas at Austin from January 30-March 27, 2021.*

Indigenous peoples are those who have creation stories, not colonization stories, about how we/they came to be in a particular place – indeed, how we/they came to be a place.
—Eve Tuck and K. Wayne Yang, 2012

A Hill to Die On

A light wind blows, bringing the smallest respite from the heat of the beating sun. I notice that my shoulders are starting to turn pink—I touch the skin and see a trail of white left by my fingertips for a moment or two, before the blood rushes back. The heat is astounding. Even having grown up in Texas, I am surprised by the dry intensity of this landscape. All around me are red rocks and blue sky. In the middle of it is a vast, gray field of stone that sticks out like a sore thumb from my vantage point on a hill, though from the highway nearby it is all but invisible. I have traveled here with the artist Will Wilson to the edge of the Navajo Nation. From the hill we stand on, one can see where borders meet between the reservation and the United States, though the land looks no different from one side to the next. I am reminded of the popular saying, "We didn't cross the border, the border crossed us." Political demarcations hold little meaning on Indigenous lands.

And we are all on Indigenous lands.

The wind picks up and I look over to see the shade on Wilson's camera flapping upwards and dangerously close to blowing away. On his left is a barbed wire fence and a steep drop down to the field of stones below: no man's land. The landscape is spread out in a seemingly impossible arrangement of vectors and planes that lend the whole area a sense of careful precision. We both know that underneath those rocks is a different story, the one that brought us here today. We are in Halchita, just inside the edges of the Navajo Nation, adjacent to the small town of Mexican Hat, Utah. The site in front of us is known as the Mexican Hat Disposal Cell and was created in 1995 as part of the Uranium Mill Tailings Remedial Action Program in the US Department of Energy. That year, the US government buried 4.4 million tons of radioactive tailings from the surrounding community (including mining waste, processing facilities, a school building, homes, and other structures infiltrated by radiation) in a pit covering sixty-eight acres in the desert (The Center for Land Use Interpretation, 2012). From Monument Valley, the nearby tourist attraction just twenty-five miles away, you'd never know it was there.

Wilson brought me here to see and photograph one of the over 500 identified abandoned uranium mines on the Navajo Nation that are in various stages of treatment by the US government. These sites, some so large they encompass the towns surrounding them, and others just holes in the ground, are remnants of an era of nuclear weapons development and testing that have left the region with a legacy of toxicity and destruction. Uranium was mined heavily throughout the Navajo Nation beginning in the early 1940s, when the United States Government sought to build and successfully detonate the world's first atomic bomb. As the radioactive material was extracted and processed by mostly Navajo miners, its particles nestled in their skin, clothing, hair, and lungs, where they would bring it back to their families. Homes built from mining refuse, known as "hot homes", slowly poisoned their inhabitants. Entire households became radioactive. Water sources that sustained whole communities were contaminated with radioactive waste. Those living on the Navajo Nation continue to experience the effects of radiation poisoning today through heightened rates of infertility, cancers, and death.

This essay traces the emergence of a series of photographs by Wilson titled *Connecting the Dots for a Just Transition* (hereafter referred to as *Connecting the Dots*) that ruminate on this toxic history. In this project, Wilson photographically surveys the hundreds of abandoned uranium mines within and along the borders of the Navajo Nation to create a visual record of the largely overlooked history of resource extraction there. Some of these places, like Mexican Hat, in Utah have been designated as "Superfund" sites by the US Environmental Protection Agency, making them among the most toxic places in the country (United States Environmental Protection Agency, 2022). I consider this legacy of extraction within the context of *Connecting the Dots* and Wilson's larger practice. Ultimately, the series is part of an ongoing effort by Wilson to respond to ecological damage on the Navajo Nation through Diné knowledge systems and cultural resilience. *Connecting the Dots* bears witness to the multi-layered trauma represented by extractive mining practices in the Southwest, and its images are a reckoning that we all must answer to.

The People's Place

The Diné, which translates simply to "the people", are known by most outsiders as the Navajo.[1]

Driving through Diné lands, one immediately senses its people's particular sense of appreciation for water. Holding tanks dot the relatively rural landscape alongside homesteads between the larger towns of Window Rock, Tuba City, Shiprock, and Kayenta, oftentimes tagged with graffiti images and words, more than one displaying the now oft-repeated phrase *Water is life*, or *tó éí iiná* in Navajo (Howe, 2021). *Water is life* is not a simile. In contrast to the idea that water sustains life, the concept that water and life are interchangeable is key to Diné ways of relating to the earth.

This grammar of equation, that the non-human is equal to its human counterpart, makes clear that human life is not only sustained but is also indelibly connected with the non-human realm in Dinétah: the Diné ancestral homelands spanning the contemporary Four Corners region covering New Mexico, Arizona, Utah, and Colorado. The current boundaries of the Navajo Nation include a portion of Dinétah, or Diné Bikéyah (meaning people's sacred lands), but only represent reduced fragments of them. Over the thousands of years that the Navajo and their ancestors have lived in Dinétah, they have articulated what art historian Janet Catherine Berlo (2009, p. 239) described as "a cogent ecological worldview concerning the spiritual and material relationships between human beings and their environment, keyed to a specific sense of place." For the Diné, place, more than just geographical space, exists as a complex construction of stories, beliefs, and memories that are re-enacted continually over time.

Wilson alludes to this interwoven sense of people and place in Diné worldviews throughout his practice. Dinétah is conceptualized as a sacred structure that physically and spiritually sustains the Diné, oftentimes equated with customary dwellings known as hogans, dome-shaped structures, whose doorway faces east towards the origins of the Navajo universe and the rising sun. In short, Dinétah is equated to home. As Berlo (2009, p. 240) has also pointed out, "While much of the Navajo landscape might strike outsiders as a tough and dramatic wilderness, European concepts such as 'wilderness' and 'landscape' make no sense in a system in which, at every level, one is within the safety of home." To poison Dinétah is therefore to poison the home and heart of the Diné.

Wilson is a citizen of the Navajo Nation who now resides in Santa Fe, NM. Having spent his formative years living in Tuba City in the Western agency of the Navajo Nation with his mother's family, Wilson intimately understands the physical, cultural, and ecological trauma that the Diné have been forced to bear due to histories of resource extraction.[2] Using vibrant coloring, deep contrasts, Wilson's photography captures vivid glimpses of a complex world that is both dazzling and haunted. His aestheticized landscapes speak to a history of landscape survey photography that began in the US in the latter half of the nineteenth century, when photographers like William Henry Jackson, Timothy H. O'Sullivan, Carleton E. Watkins, and others poured westward to create images of landscapes for visual consumption and economic exploitation. These photographs often erased the Indigenous inhabitants of these spaces, giving the impression that lands were uninhabited and available for taking.

In more recent continuations of this tradition, photographers have turned to devices such as helicopters, planes, and drones to transport cameras high above the land to capture birds-eye perspectives that imbue landscapes with a sense of detached surveillance. *Connecting the Dots* cites these various traditions of survey photography while also pointing to the detrimental ecological practices they precipitated.

Mexican Hat Disposal Cell Redux, Halchita, Utah, Navajo Nation (2019) encapsulates these goals through its wide-ranging birds-eye view of the remediated Superfund site. The photograph was captured by Wilson using a drone, one of a small fleet he has collected in pursuing this project. A large hexagonal formation with outgrowths inhabits the foreground of the scene. Though from this distance it appears mostly smooth, it is in fact made from multiple layers of stones arranged upwards to a flat vertex. The shape looks alien in the serpentine landscape of cliffs, rock formations, and canyons surrounding it. This is the Mexican Hat Disposal Cell, where the waste of the nearby Halchita uranium processing facility and community is buried. In the middle distance is the contemporary town of Halchita. Beyond that, along the horizon is the iconic Monument Valley, featured most famously in the films *Stagecoach* (1939), *The Searchers* (1956), and *Forrest Gump* (1994). Popular culture has transformed the area into a mythic representation of the American West and the Western genre, erasing the calamity that exists just down the road. To understand this rupture in remembrance, one must look to what is buried in the Mexican Hat Disposal Cell: uranium.

The Canaries in the Uranium Mine

The world's first atomic weapon was tested in the United States on July 16, 1945, at the Alamogordo Bombing Range (in the Jornada del Muerto desert), just fifteen miles by road to the Southeast border of the Mescalero Apache reservation in Southern New Mexico. J. Robert Oppenheimer, the lead physicist of The Manhattan Project who oversaw the weapon's construction and detonation, named the test "Trinity."[3] The atomic bomb (dubbed "Gadget" by its makers) was detonated less than three years after President Franklin Roosevelt first subsidized the project in 1942 (Kelly, 2007; Rhodes, 1986; Sullivan, 2016).[4] The US had declared war on Japan the year before, bringing World War II directly to its shores and catalyzing the ongoing attempts to produce an atomic weapon before their enemies did. Physicists in the Manhattan Project focused on the ability of uranium to catalyze a massive atomic energy release. Uranium mining (and mining for its cousin vanadium) therefore began in earnest in the Colorado Plateau during the 1940s to provide the materials necessary to power these new atomic weapons. The Navajo Nation, located in the southern portion of this region, was identified as a primary potential source of the material due to the massive and untouched stores of uranium on their land.

Uranium is known as *leetso* or "yellow dirt" to the Diné, and they have long described it in their stories as an evil matter to be left undisturbed. Anna Rondon (Diné), a member of the Southwest Indigenous Uranium Forum, recounted at the World Uranium Hearing in Salzburg in 1992 that the yellow substance known as uranium was never intended to be brought up from the underworld. She elaborated:

…as Navajo people, we were given the choice when we came from the underworld, the fourth world, we came to the emergence of this world that we are living now, the glittering world. The Creator told us that we have a choice, a choice to use the corn pollen, which is a yellow substance that we use. It contains the positiveness of life. We were also given a choice to use the yellow cake dirt which we were told by the Creator contained the negative particles of life. So we had a choice and we chose the corn pollen, which is the beauty way. We chose that for all people (Rondon, 1993, p.157).

Under pressure from the US government and with the false understanding that vanadium was actually the sought after material, not uranium, by the late 1940s the Navajo Nation opened up to mining by external companies, given that the Diné were prioritized for jobs in mining operations. Though the growing sector of uranium mining did provide some level of economic stability for families following a devastating period of livestock reduction in the 1930s, it came at a severe health risk for both the miners and their families.[5] Studies in Europe as early as 1879 noted increased cases of lung cancer in uranium miners, though the link between radiation poisoning and cancer would not be fully developed until the mid-twentieth century, and even then, miners on the Navajo Nation were not informed of the dangers (Axelson, 1995, p. 37). Their bodies, however, registered the damage of radiation poisoning within a matter of years. As Wilson once said with painful accuracy, "We're like the canaries in a coal mine" (Jadrnak, 2016).[6]

The full effects of long-term radiation poisoning from uranium exposure wouldn't be pieced together until much later, however. Traci Brynne Voyles described the health risks or uranium mining in detail in her book *Wastelanding: Legacies of Uranium Mining in Navajo Country*:

By the mid-1980s, researchers found astronomical rates of cancer deaths among former uranium miners. Miners contracted lung cancer at rates fifty six times higher than the national average, and had an average life expectancy of only forty six years. Rates for stomach cancer were eighty two times the national average. Miners were more than 200 times more likely to get liver cancer, almost fifty times more likely to get prostate cancer, and over 60 percent more likely to have cancers of the bladder or pancreas. Nor were cancers the only health problems among former miners and their families: researchers also found increased incidents of tuberculosis, fibrosis, silicosis, and birth defects, all linked to exposure to uranium from mines and mills. Radiation-related diseases are now endemic to many parts of the Navajo Nation, claiming the health and lives of former miners to be sure but also those of Navajos who would never see the inside of a mine. Diné children have a rate of testicular and ovarian cancer fifteen times the national average, and a fatal neurological disease called Navajo neuropathy has been closely linked to ingesting uranium-contaminated water during pregnancy. Studies have also found that uranium has genotoxic and mutagenic effects; that is, uranium poisoning can change the genetic material of a chronically exposed population, even further expanding uranium's influence on future populations in ways that are yet unknown (Voyles, 2015, p. 4).

Voyles went on to argue for understanding uranium mining on the Navajo Nation as a case of environmental racism, using the term *wastelanding* to describe a settler colonial practice that "involves a deeply complex construction of that land as either always already belonging to the settler— his manifest destiny—or as undesirable, unproductive, or unappealing..."

(Voyles, 2015, p. 7). A central component of *wastelanding* is the erasure of Indigenous relationships to land, and of Indigenous peoples themselves. Wilson's project attempts to reverse this *wastelanding* process in the ways that it re-visualizes the effects of uranium mining on Native lands, making the slow-motion devastation of Dinétah present and urgent for viewers of the project. In doing so, he delineates landscapes understood in a Western sense as passive as instead being part of a larger Diné system, in which people and place are indelibly intertwined.

Uranium extraction was fully outlawed on the reservation in 2005 following years of activism by environmental justice organizations (Voyles, 2015, p. xv). But the damage had been done. The US Environmental Protection Agency has estimated that between the years 1944 to 1986 alone, thirty million tons of uranium ore were extracted from Diné lands. Today the Navajo Nation contains 523 identified abandoned uranium mines (though there are thought to be many more) that continue to emit higher than normal levels of radiation into surrounding areas (United States Environmental Protection Agency, 2022). Though some of these sites have been remediated by the EPA, meaning that radioactive tailings and other mining debris have been buried in lined pits and covered, the majority of these mines sit open and are poorly marked, if at all, and continue to emit toxic contaminants into the surrounding environment (Voyles, 2015, p. 3). Long-term exposure to radiation has resulted in painful losses to Diné lives and well-being, highlighting the continued dangers of uranium's mining in Diné lands.

The Emergence of *Connecting the Dots*

In the late 1970s and 1980s on the Navajo Nation, uranium's presence remained ubiquitous, even as it grew increasingly absent. Wilson spent his childhood playing with other children at the abandoned Tuba City Rare Metals Mill site just outside of town. Its crumbling detritus was the playground and meeting place for local youths for over twenty years. The uranium processing facility had been left to mostly disintegrate after its closure in 1966. In the ten years the mill operated, it processed around 800,000 tons of uranium ore (US Department of Energy, 2021). Between 1966 and 1988 when site remediation began, radioactive tailings from the mill's operations were left to blow from evaporation ponds at the site into the nearby town and water sources. Wilson's photograph *Rare Metals Disposal Cell, Tuba City, Arizona, Navajo Nation* (2020) visualizes the site as it is today, the eerie cousin of *Mexican Hat Disposal Cell Redux, Halchita, Utah, Navajo Nation*. The former's format is essentially the same- a stone formation in the foreground (this one vaguely pentagonal), surrounded by an arid landscape and natural formations, all photographed with a drone from a birds-eye view. On the right side of the photo, buttressing the road, one can make out the skeletons of the homes where mill workers previously lived. Of central importance to Wilson is the proximity of the Rare Metals Disposal Cell to the Moenkopi Wash (seen as a snaking canyon along the top edge of the scene), which contains a tributary of the Little Colorado River. In Wilson's own words, "A major concern is that contamination will migrate into the Moenkopi Wash, a vital water source that irrigates the cornfields of Curley Valley below Tuba City. Resident Rose Williams—a farmer and activist—cultivates non-GMO corn along the Moenkopi Wash that is the source of the Taada'diin, or sacred corn pollen... She currently lives about a mile from the Rare Metals site. My grandfather also kept cornfields in this foodshed, now threatened by toxic desecration" (Wilson, 2021, p. 7).

As Wilson points out, even buried radioactive materials are still radioactive, and they pose a threat to nearby water sources and those who depend on them. During the monsoon season in the Southwest, typically June through September, heavy rains often flood valleys such as the one where Rare Metals is located. Such flooding seeps into the ground, stressing the lining of remediated pits through their annual cycles of saturation and desaturation. And though the design of these sites is intended to force water runoff from the pit itself, water inevitably seeps into these radioactive burial sites and infiltrates the groundwater, transporting traces of radioactive waste with them and to the citizens of the Navajo Nation.

Shiprock Disposal Cell, Shiprock, New Mexico, Navajo Nation (2020) visualizes this relationship between mining sites and water sources even more dramatically. The town's namesake, Shiprock, is visible in the distance of Wilson's photograph, highlighted against the horizon line by late afternoon light and storm clouds. The disposal cell at this site occupies the middle ground of the scene. It is perched precariously along the edge of the San Juan River. In an even more harrowing layout than the previous *Connecting the Dots* photographs discussed, the town of Shiprock butts up directly against the Shiprock disposal cell. *Shiprock Disposal Cell, Shiprock, New Mexico, Navajo Nation* reads as the most urgent of Wilson's images for this reason- here, the physical proximity of radioactive waste to the water sources and the people depending on them is particularly distressing.

In one of the earliest photographs from the *Connecting the Dots* series titled *Auto Immune Response, Mexican Hat Disposal Cell, Halchita, Utah, Navajo Nation* (2019), two figures appear at opposite ends of an extended panorama overlooking Mexican Hat. The image overlaps with an earlier series known as *Auto Immune Response (AIR)* (2004–ongoing) through its use of a performative figure (played by Wilson) known simply as the protagonist. In *Auto Immune Response, Mexican Hat Disposal Cell, Halchita, Utah, Navajo Nation* two protagonists appear on the hill in Halchita overlooking the Mexican Hat Disposal, the same hill Wilson and I stood on, and look out onto the sea of stones below. The left protagonist guides a drone out over the scene as his companion watches on. This photograph is unique in that it is formed from multiple images and photographic processes spliced together; the black and white portions are taken using wet collodion photography, while the color portions are edited in from digital photographs. Much of the early landscape survey photography of the 1860s-1870s was taken using the wet collodion method (and the later, closely related dry collodion method). Wilson's use of it here thereby ties him to earlier photographic histories and techniques even while he subverts it through his digital additions.

There is nothing immediately alarming about the landscape Wilson presents in *Auto Immune Response, Mexican Hat Disposal Cell, Halchita, Utah, Navajo Nation*. Upon closer inspection, however, viewers may note that the protagonists don gas masks. It is not a coincidence that increased use of gas masks in WWII overlapped with the development of the atomic bomb and the push to mine uranium on Native lands; these were corollaries of the same trajectory towards increasingly brutal forms of chemical warfare. The gas mask can also be tied to a particular history of contemporary Native

American art in which the mask has become a sign of the widespread poisoning of Indigenous lands and bodies.[7] This poisoning is oftentimes both literal and metaphorical, referring to both the compromised ecological balance of the land and people, as well as to the resulting oppression of Indigenous ways of being. By integrating the gas mask into his work, Wilson delineates these interconnected forms of physical, ecological, and cultural violence—what Elizabeth Cook-Lynn explains as the process of erasure in a settler colonial world—and highlights them as interwoven parts of a larger system of global Indigenous erasure (Cook-Lynn, 2007).

The Beginning and the End

The *AIR* series converges with the newer *Connecting the Dots* project in the same cultural, environmental, and geographic sphere of Dinétah. Both series advocate a return to balance in the irradiated lands of Dinétah. Where *AIR* emerges from Wilson's individual wanderings, *Connecting the Dots*, is inherently community-centered and interdisciplinary, with Wilson's long-term goal being to photograph, catalog, and map each of the more than five hundred uranium mines within Dinétah. The project is therefore visual, textual, and geographical in scope, bringing together visual arts and sciences to strengthen the case for restorative justice for the Diné. Beyond creating images to bring awareness to the little-known reality of uranium extraction in the Southwest, *Connecting the Dots* has evolved to include multiple elements of community outreach within the Navajo Nation. Wilson partnered with the Diné College in Tsaile, AZ on the Navajo Nation to create a "Reframing Indigenous Remediation" speaker series and a website (currently being built) that brings widespread awareness to this extractive history. Ultimately, Wilson's goal is to empower Diné citizens to move into scientific and cultural fields to be advocates for Indigenous-led remediation efforts.

Back on that hill overlooking Halchita, the breeze carried with it a myriad of stories of past, present, and future. Stories of the ancestors. Stories of the people. Stories of the monsters.[8] Each gust of wind on that hill carried complex stories of ecological entanglement that require tuning into. The human and non-human world are emphatically interwoven in Dinétah.[9]

Imbalance in one leads to imbalance in the other. Resource extraction is one imbalance among many that the Diné navigate in their daily existence. For those living on the Navajo Nation and in Indigenous communities worldwide, Wilson presents a familiar tragedy of destruction. Yet, as Wilson makes clear, this tragedy is far from settled, and new stories are still being written.

Notes

[1] The term Navajo has Spanish colonial origins and is still often used interchangeably with Diné to identify the people of the Navajo Nation. Though I lean more heavily on Diné as the preferred self-identifier, I use both terms to reflect the various ways that this group is recognized and recognizes themselves in various settings.

[2] The Navajo Nation is split up into five agencies with various chapters and communities spread between them. These agencies are the Eastern, Shiprock, Chinle, Fort Defiance, and Western. To'Nanees'Dizi (otherwise known as the Tuba City chapter) is in the Western Agency portion of the reservation that lies within Arizona state boundaries.

[3] Oppenheimer purportedly named the Trinity test after a poem by John Donne, "Holy Sonnet XIV: Batter My Heart, Three-Personed God."

[4] The impetus for The Manhattan Project began in 1938 when President Roosevelt received a letter from Albert Einstein and Leo Slizard warning him of the Germany's progressing attempts to build an atomic bomb for Nazi use. Roosevelt responded by forming The Advisory Committee on Uranium in 1939, which would eventually lead to the creation of The Manhattan Project. For more, see: Richard Rhodes, *The Making of the Atomic Bomb*, New York: Simon and Schuster, 1986; *The Manhattan Project: the Birth of the Atomic Bomb in the Words of Its Creators, Eyewitnesses, and Historians*, ed. by Cynthia C. Kelly, New York: Black Dog & Leventhal Publishers, 2007; Neil J. Sullivan, *The Prometheus Bomb: The Manhattan Project and Government in the Dark*. Dulles: Potomac Books, 2016.

[5] Overseen by Bureau of Indian Affairs Commissioner John Collier, Navajo stock reduction aimed to cut the reservation's goat, horse, and sheep populations by nearly half to confront the land's growing desertification from overgrazing and environmental changes on the reservation. In Diné thought, sheep are a gift from the Diyin Dine'é (The Holy People) and are meant to be life-sustaining in physical, economic, and spiritual ways. Women are the primary sheep keepers of families and clans and have therefore historically held a significant amount of power in Diné society. The Diné continue to face economic and cultural challenges resulting from the livestock reduction program, which not only devastated an essential source of food and economic value for families, but also undermined the centrality of sheep and women herders in Diné culture. For more, see: Marsha Weisiger, *Dreaming of Sheep in Navajo Country* (Seattle and London: University of Washington Press, 2009).

[6] This is a reference to the popular adage of the "canary in the coal mine." Beginning in 1911, canaries were used in coal mines to detect the presence of carbon monoxide and other harmful gases. If a canary died in a mine, it was a sign to the workers to evacuate immediately.

[7] Other contemporary Native American artists to appropriate the gas mask include Naomi Bebo (Menominee/Ho-Chunk), Gregg Deal (Pyramid Lake Paiute), Bunky Echo-Hawk (Pawnee/Yakama), Dallin Maybee (Seneca and Northern Arapaho), Da-ka-xeen Mehner (Tlingit/N'ishga), and Virgil Ortiz (Cochiti Pueblo), among others.

[8] The Diné Bahane' (Navajo creation story) describes how two brothers known as the Hero Twins, Naayéé' Neezghání (Monster Slayer) and Tó Bájíshchíní (Born for Water), rid Diné Bikéyah of the monsters known as Naayéé' that were terrorizing their world. Others have noted the parallels between the Hero Twins' defeat of the monster Yé'iitsoh (also referred to as Yeetso, meaning Big Monster) and leetso, the Navajo word for uranium. Wilson specifically referenced the Hero Twins and other stories from the Diné Bahane' throughout his *Auto Immune Response* project. See: Esther Yazzie-Lewis and Jim Zion, "*Leetso*, The Powerful Yellow Monster: A Navajo Cultural Interpretation of Uranium Mining," in *The Navajo People and Uranium Mining*, ed. Brugge et. al. (Albuquerque: University of New Mexico Press, 2006).; Paul G. Zolbrod, *Diné bahane': The Navajo Creation Story* (Albuquerque: University of New Mexico Press, 1984).

[9] The human and non-human realms are often collectively referred to today as other-than-human, or more-than-human. These are two terms that have risen in scholarly usage in the 21st century to de-center the human/non-human dichotomy and prioritize the interrelationships of all beings. While they have been embraced across a growing number of (mostly humanities) fields, the concept of co-constitution and entangled human/non-human relationships has long been inherent to Indigenous epistemologies worldwide. See: David Abram, *The Spell of the Sensuous: Perception and Language in a More-than-Human World* (New York: Vintage Books, 1996).; Joni Adamson and Salma Monani, eds., *Ecocriticism and Indigenous Studies: Conversations from Earth to Cosmos* (Abingdon: Routledge, 2016).; Paul Berne Burow et. al., "Unsettling the Land: Indigeneity, Ontology, and Hybridity in Settler Colonialism," *Environment and Society* 9, no. 1 (2018).: 57–74; Robin Wall Kimmerer, *Braiding Sweetgrass: Indigenous Wisdom, Scientific Knowledge and the Teachings of Plants* (Minneapolis: Milkweed Press, 2014).

References

Axelson, O. (1995). Cancer Risks from Exposure to Radon in Homes. *Environmental Health Perspectives*, 103 (2), p. 37. Available at: http://www.ncbi.nlm.nih.gov/pubmed/7614945

Berlo, J. (2009). Alberta Thomas, Navajo Pictorial Arts, and Ecocrisis in Dinétah. In Braddock, A. et. al. (Eds.), *A Keener Perception: Ecocritical Studies in American Art History*. Tuscaloosa: The University of Alabama Press, pp. 239-240.

The Center for Land Use Interpretation. (2012). *Land Use Database.* Available at: https://clui.org/ludb/site/mexican-hat-uranium-disposal-cell.

Cook-Lynn, E. (2007). *Anti-Indianism in Modern America: A Voice from Tatekeya's Earth*. Urbana and Chicago: University of Illinois Press.

Howe, C. (2021). *Bringing Clean, Running Water to the Navajo Nation*. Available at: https://www.rwjf.org/en/blog/2021/05/bringing-water-to-the-navajo-nation.html

Jadrnak, J. (2016). Artist series imagines photographer as sole survivor. *Albuquerque Journal*, December 30. Available at: https://www.abqjournal.com/917933/canaries-in-a-coal-mine.html

Kelly, C. (Ed.). (2020). *The Manhattan Project: the Birth of the Atomic Bomb in the Words of Its Creators, Eyewitnesses, and Historian*. New York: Black Dog & Leventhal Publishers.

Rhodes, R. (1986). *The Making of the Atomic Bomb*. New York: Simon and Schuster.

Rondon, A. (1993). Poison Fire, Sacred Earth: Testimonies, Lectures, Conclusions, The World Uranium Hearing, Salzburg. In Nahr, S. (Ed.), *World Uranium Hearing 1992*, pp.157-158, München.

Sullivan, N. (2016). *The Prometheus Bomb: The Manhattan Project and Government in the Dark*. Dulles: Potomac Books.

US Department of Energy. (2021). *Tuba City, Arizona, Disposal Site Fact Sheet*. Available at: https://www.energy.gov/sites/default/files/2021-09/TubaCityFactSheet.pdf

United States Environmental Protection Agency. (2022) *What is Superfund?*
Available at: https://www.epa.gov/superfund/what-superfund

United States Environmental Protection Agency (2022). *Navajo Nation:
Cleaning Up Abandoned Uranium Mines*. Available at:
https://www.epa.gov/navajo-nation-uranium-cleanup

Voyles, T. V. (2015). *Wastelanding: Legacies of Uranium Mining in Navajo
Country*, pp. xv, Minneapolis: University of Minnesota Press.

Wilson, W. (2021). Will Wilson: AIR/Survey. In Schedeen, K (Ed.) *Will Wilson:
AIR / Survey*, Austin: Visual Arts Center, University of Texas at Austin.

Connecting the Dots: Will Wilson

Figure 9.1
Mexican Hat Disposal Cell Redux, Halchita, Utah, Navajo Nation, 2019.
Archival pigment print. Size variable.

Figure 9.2
Mexican Hat Disposal Cell, Detail 1, Halchita, Utah, Navajo Nation, 2019.
Archival pigment print. Size variable.

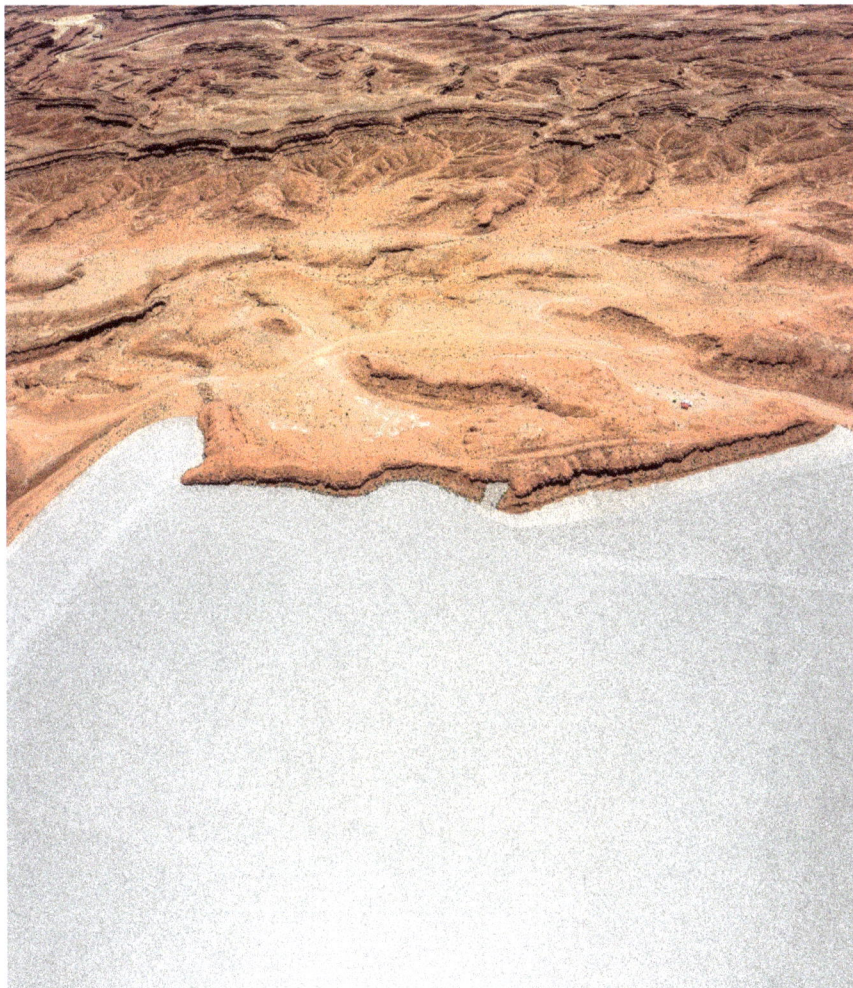

Figure 9.3
Rare Metals Disposal Cell, Tuba City, Arizona, Navajo Nation, 2020.
Archival pigment print. Size variable.

Figure 9.4
Shiprock Disposal Cell, Shiprock, New Mexico, Navajo Nation, 2020.
Archival pigment print. Size variable.

Figure 9.5

Auto Immune Response, Mexican Hat Disposal Cell, Halchita, Utah, Navajo Nation, 2019.

Digitype (archival pigment print from original tintypes and digital captures).

Size variable.

Chapter 10

Regenerative Relationships of Harvesting Clay for Healthy Soils and Communities in the Shinnecock Nation

Kelsey Leonard

Indigenous communities across the world have faced significant challenges to maintaining ancestral practices and knowledge related to clay harvesting. However, there have also been significant efforts by Indigenous communities to resist the impact of colonialism and to revitalize and promote clay practices. As described in this chapter, Indigenous Peoples are rematriating practices to steward past, present, and future clay relationality. On Long Island, New York, it is common for pottery shards to be uncovered when the Land[1] is disturbed by homeowners or infrastructure projects. Within each shard is a history of Indigenous coastal clay practice that spans centuries, if not millennia. As a climate and Water scientist at the intersection of the human-environment nexus, I have a persistent concern as to whether the increasing impacts of climate change will irrevocably shift Indigenous access to clay in our ancestral territory of Paumanok (Long Island).

The Shinnecock Nation, located on the eastern end of Long Island, New York, has a storied history of clay harvesting. These clay harvesting practices have been developed over millennia of observation and experimentation with the Lands and Waters. More recently, many of the founding scholars of geology built their careers exploring the geology of Long Island. In 1842, William Mather created a geological map of Long Island identifying topography inclusive of soil types (Fig. 10.1). In 1883, Frederick Jams Hamilton Merrill published a paper that identified many of the clay features that are unique to the island's ecosystems.

Figure 10.1 *William Mather, Geological map of Long & Staten islands with the environs of New York, 1842. Lionel Pincus and Princess Firyal Map Division, The New York Public Library.*

As the centuries progressed, settler-colonial powers would extract this clay for brick production and other ceramic wares necessary to meet colonial desires and expansion (Merrill, 1883). However, less documented in the written record are stories of Shinnecock clay relationality. The Shinnecock have a deep connection to the Land, the Water, and the surrounding environment, and our ancestral practices for harvesting and using clay are guided by principles of sustainability and stewardship – namely, not to overharvest and only take what we need in recognition of the gift being provided for our sustenance. These practices are deeply rooted in cultural traditions and often have spiritual and ceremonial significance. The Shinnecock have a long history of using clay for a variety of purposes such as pottery, construction, and art. These practices linked to ancestral methods of harvesting clay are not only important for cultural continuity but important for promoting healthy soils and communities.

The term "soil" in relation to "clay" requires further elaboration to understand their connection and implications. While clay is a type of soil, it is important to recognize that soil is a complex mixture of organic matter, minerals, liquids, and organisms. Different types of soils, such as sandy, silty, peaty, chalky, and loamy soil, and clay possess distinct chemical and physical properties that make them suitable for various applications. The health of the soil and the degradation of soil deposits can significantly impact the characteristics and workability of clay. Degradation of soil, such as compaction or loss of organic matter, decreases the plasticity or malleability of clay, making it challenging to employ traditional techniques like coiling. This can affect the bonding and forming of clay particles. Consequently, current-day practices by Shinnecock harvesters involve the continued acquisition of knowledge to identify and manage newly located clay deposits that may have been affected by such degradation processes.

Moreover, compaction of the Land, a common issue associated with soil, can have broader implications beyond clay workability. Compacted soil reduces the total pore space available, affecting plant growth and root development. When the soil becomes compacted, it becomes harder for roots to penetrate and grow, limiting access to essential nutrients and Water. Additionally, compacted soil contributes to issues with runoff, as Water does not penetrate the soil profile but instead runs along the surface. This can lead to problems with drainage, erosion, and contamination. Understanding these soil-related challenges is vital for Land management practices and preserving healthy soils that support clay harvesting and overall ecosystem health.

Overharvesting clay on Long Island can deplete local deposits, making accessing high-quality clay for cultural practices increasingly challenging. Unregulated or unsustainable clay extraction leads to the degradation of clay-rich areas, disrupting soil ecosystems and decreasing the availability of clay resources. These factors emphasize the need for sustainable clay harvesting practices, responsible Land management, and the implementation of measures to mitigate and adapt to the impacts of climate change. The 2019 Shinnecock Indian Nation Climate Vulnerability Assessment and Action Plan outlines the Nation's goals stating, "Because of the strong ancestral connection to the land and water, as well as a commitment to promoting environmental goals to protect human health, the Nation values its natural resources. Climate change is viewed as a threat to aquatic water quality and habitat, especially traditional fisheries such as shellfish, and to the *health of the soil*, groundwater, and plants" (p. 12, emphasis added). Understanding the connections between soil, clay, climate change, and overharvesting is crucial for promoting sustainable practices, protecting soil health, and ensuring the availability of clay resources for future generations on Paumanok (Long Island). By addressing these challenges through conscious Land use, conservation efforts, and climate resilience strategies, we can support the preservation of clay harvesting traditions while safeguarding the region's ecological integrity.

The Shinnecock have a deep understanding of the interconnections between human activities and the natural environment. We recognize that healthy soils are essential for the health and well-being of our communities, as well as for the health of the broader ecosystem. Our Indigenous practices for harvesting and using clay often involve a holistic approach to Land and Water use and management, which recognizes the importance of preserving soil health and promoting biodiversity. One of the key features of Shinnecock clay harvesting is the use of selective harvesting. Selective harvesting involves taking only the amount of clay that is necessary, leaving behind healthy soil and vegetation. This approach ensures that the ecosystem is not overexploited and that the surrounding environment remains healthy and sustainable. Another important aspect of Shinnecock clay harvesting practices is the use of natural materials and methods, such as the use of wood or seaweed to create fires for pit firing, an ancestral method of firing clay. This approach avoids the use of heavy machinery and equipment, which can damage soil structure and contribute to soil degradation. Pit firing also allows for the replenishment of nutrients and organic matter into the soil, which promotes healthy soil and biodiversity.

One of the primary uses for clay among the Shinnecock was for the creation of pottery. Shinnecock pottery is often made using a technique known as *coiling*, which involves rolling the clay into long, thin ropes and then stacking and coiling them to form the shape of the pot. The pots were then smoothed and decorated using traditional techniques such as incising and appliqué. In addition to pottery, clay was also used for a variety of other purposes among the Shinnecock people. This included the creation of clay pipes for smoking and ceremonial activities (National Museum of the American Indian, n.d.).

The Shinnecock have often used shells to temper our clay, a technique that enhances the firing process and preserves the integrity of the pottery, drawing on our deep knowledge of local clay deposits and the characteristics of different areas. However, the future of this practice may be threatened due to the increasing acidification of bay and Ocean Waters resulting from climate change and anthropogenic Water pollution. These environmental factors have led to a decline in shellfish populations within local Waters, posing a challenge to the traditional techniques and cultural practices of the Shinnecock community. Preserving the vitality of these practices will necessitate a commitment to Land and Water sustainability, addressing the root causes of Water pollution and mitigating the impacts of climate change to safeguard the essential resources that contribute to the unique character of Shinnecock pottery.

Shinnecock clay harvesters have a deep understanding of the properties of clay and its uses. Shinnecock clay harvesting practices go beyond promoting healthy soils and play an important role in maintaining cultural ways of knowing and being. However, despite the importance of these practices, they have been threatened by the expansion of modern industrial techniques. Large-scale mining or excavation of clay can cause significant damage to soil health and biodiversity, leading to soil degradation, habitat destruction, and the loss of cultural heritage. The quality of clay used in pottery is highly dependent on the health of the soil from which it is extracted. Soil depletion and erosion caused by deforestation, intensive agriculture, and other human activities can lead to a decrease in the availability of high-quality clay, as well as a decrease in the fertility of the soil. On Long Island, these harmful activities are linked to increased development and island sprawl, claiming ecologically sensitive pine barrens and increasing wastewater runoff into local waterways, alongside a continued legacy of agricultural production from potato farms to vineyards. This can affect the quality and quantity of pottery produced, as well as the sustainability of the practice.

In recent years, there has been a growing movement towards the recognition and revitalization of Shinnecock harvesting practices. Shinnecock harvesters have been advocating for the rematriation of our traditional Lands and Waters, which includes the promotion of sustainable Shinnecock methods of harvesting and using clay. This movement also aligns with the broader emergence of ceramic ecology.

Ceramic ecology refers to an interdisciplinary approach that combines ceramics, ecology, and sustainability. It explores the intersection of ceramics as an artistic medium with ecological principles, environmental consciousness, and sustainable practices. Ceramic ecology aims to promote a holistic understanding of the environmental impact of ceramic art and to encourage responsible and ecologically conscious approaches to ceramic production, use, and disposal. Ceramic ecology emphasizes the importance of responsible sourcing of materials used in ceramics, such as clay, glazes, and other additives. Thus, factors like the environmental impact of mining or extracting clay, the use of sustainable materials, and the conservation of finite resources are also considered. It promotes awareness and understanding of the ecological context in which ceramics exist. Artists are encouraged to consider the broader environmental implications of their work and explore themes related to nature, climate change, biodiversity, and human-nature relationships. In ceramic ecology the entire life cycle of ceramic objects is considered, from raw material extraction to their eventual disposal. This includes evaluating the environmental impact of production, transportation, and use as well as the potential for recycling, reusing, or repurposing ceramics at the end of their life.

This chapter aims to identify a definition of Indigenous ceramic ecology in relation to Shinnecock rematriation of clay harvesting practices. In exploring the topic of clay harvesting practices in the Shinnecock Nation, it is crucial to understand these practices' historical evolution, their cultural significance, and their impact on soil health and community well-being. By delving into the techniques and processes involved in clay harvesting, we can uncover ancestral wisdom that has guided these practices for generations. Next, this chapter examines the contemporary challenges and opportunities that shape the present state of Shinnecock clay harvesting as well as the initiatives being undertaken to address these issues. Finally, the chapter explores the barriers to the future rematriation of regenerative clay harvesting practices to reveal the steps needed to build a sustainable and culturally resilient future for clay and the Shinnecock.

Historical Overview of Indigenous Clay Harvesting Practices in the Shinnecock Nation

The impact of colonialism on Indigenous clay harvesting practices has been significant, leading to the loss of traditional knowledge and practices, as well as the exploitation of natural resources and degradation of the environment. The arrival of European colonizers to Long Island brought about significant changes in Land use and management, as well as the displacement and marginalization of Indigenous communities, including the Shinnecock Nation. The geology of Long Island, including its unique clay deposits, has been the topic of scientific study for centuries. Newberry (1895) discussed the prevalence of Amboy clays on Long Island, noting that the deposits extend as far east as Montauk and reappear in the Aquinnah Cliffs (Gay Head Cliffs) of Noepe (Martha's Vineyard). Clay is a key relation embedded within the creation stories of Indigenous Peoples of the Atlantic shores. Its physical presence is also a contemporary reminder of the ancestral connections of eastern Algonkian nations. However, continued Indigenous stewardship of these clay sites has faced many challenges throughout history due to settler-colonialism and Land alienation.

One of the primary impacts of colonialism on clay harvesting practices was the introduction of capitalist economic systems, which placed a premium on the extraction and exploitation of natural resources. Many of the unique clay types of the marine coastal environments were named after European settler-colonialists such as Gardiner Clay, named for Lion Gardiner who created the first English colony in what is currently known as New York (Landsman, 2011; Bokuniewicz, 1980). In addition to the economic and environmental impacts of colonialism on clay harvesting practices, there were also significant cultural impacts. Indigenous knowledge and practices were often devalued or suppressed by European colonizers, leading to decreasing knowledge among the Shinnecock of practices related to clay harvesting and use. However, at the same time that Indigenous practices were being suppressed, settler-colonial collectors of Shinnecock material culture were exhibiting objects taken from the community. For example, an exhibition featuring the collection of William Wallace Tooker in 1896 was front page news in the *Long Islander* and notably included the following statement regarding Shinnecock pottery: "One earthen vessel, fifteen inches deep, thirteen inches in diameter at the top and tapering like a bee hive toward the apex, which is complete, the fragments of the same having been all gathered by Mr. Tooker and cemented together, is believed to be the only perfect specimen of Indian pottery discovered on Long Island" (*Long Islander*, 1896, p. 1). Today, the impact of colonialism on clay harvesting practices continues to be felt by Shinnecock makers, many of whom are working to revitalize and promote our ancestral practices despite much of ancient Shinnecock pottery remaining in the hands of museum and private collectors. Efforts are underway to rematriate Shinnecock artifacts, knowledge and practices related to clay harvesting and use, and to promote the recognition and preservation of Shinnecock cultural heritage.

Long Island also hosted commercial pottery production from 1804 to the early 1900s. Huntington Pottery was the largest and longest running commercial pottery business on Long Island. Pottery production at this time included redware and stoneware. Smith describes these pottery pieces as the "Tupperware" of the time and its mass production meant that Long Island clay was highly sought after (Smith, 2022). The mass production and widespread use of pottery made from Long Island clay underscore its high demand and value during that period. However, over the centuries, the availability of Long Island clay has diminished, turning it into a scarcer resource. This scarcity has implications for the Shinnecock Nation's pottery practices, as it presents challenges in accessing the necessary clay for our cultural and artistic expressions.

Despite the impact of colonialism, the Shinnecock have maintained a connection to our ancestral Lands and Waters for harvesting and using clay. In recent years, there has been a renewed interest in Shinnecock clay harvesting practices, as well as a growing awareness of their environmental and cultural benefits. Efforts to promote the resurgence of Shinnecock clay harvesting practices have included the use of educational and outreach programs, as well as art science communication – a multidisciplinary approach that weaves together artistic expressions and science to foster dialogue on the human-environment nexus. These efforts have sought to promote a deeper understanding of the cultural and environmental significance of these practices and support the transmission of this knowledge to future generations.

At the heart of these efforts is a commitment to promoting sustainable Land and Water governance. Shinnecock clay harvesting practices are often guided by principles of sustainability and stewardship, which prioritize the health and well-being of the Land, Water and community. These practices have the potential to offer valuable insights into sustainable environmental governance and policy and can serve as models for promoting sustainability more broadly.

Cultural Significance of Clay Harvesting Practices in the Shinnecock Nation

For the Shinnecock Nation, clay harvesting practices have played a significant role in our cultural traditions for generations. Clay has been used for a wide range of purposes, including creating pottery, jewelry, and ceremonial objects. The Shinnecock have a deep connection to the Land and Water and to the materials that are found within it, and this connection is reflected in our practices of clay harvesting and use. One important aspect is the recognition of the spiritual significance of clay. Clay is seen as a gift from the Creator and is treated with respect and reverence.

The process of clay harvesting involves careful observation in which harvesters look for signs of the presence of clay, such as the presence of certain plant species, and use traditional techniques to excavate the clay from the earth. Once the clay has been harvested, it is prepared for use through a process of cleaning and refinement. Pottery and other objects made from clay are often decorated with intricate designs that reflect Shinnecock cultural traditions and beliefs. These designs can include representations of animals, plants, and other elements of the natural world, as well as symbols that represent important cultural concepts and ideas. These objects are often used in ceremonial contexts, such as weddings, funerals and other important life events, and are considered to be important expressions of Shinnecock cultural heritage.

Clay is formed from the earth itself, and the process of creating objects from clay involves a deep engagement with the natural environment. The harvesting and preparation of clay require careful observation of the Land, and the creation of objects from clay is seen as a way of bringing the human and natural world together in harmony and balance. The meaning of clay objects can also be deeply personal and individual, reflecting the experiences and perspectives of the creator. For example, a potter might create a clay vessel that reflects their own unique vision of the world, incorporating personal symbolism and meaning into the design. This individual expression of creativity is an important part of the cultural significance of clay objects. In addition to their cultural and symbolic meanings, clay objects also have practical uses. Pottery, for example, can be used for cooking, storage, and transportation of food, medicines, Water, and other items, reflecting the importance of practical considerations in the creation of these objects. The combination of practical and symbolic uses reflects the deep connection between cultural traditions and daily life, and the way in which cultural practices are woven into the fabric of everyday existence. The creation of clay objects is an important expression of Shinnecock cultural heritage and identity, reflecting the diversity and complexity of our human experience.

In Shinnecock culture, clay harvesting practices are deeply interconnected with other cultural practices, reflecting interdependent relationships. The harvesting and preparation of clay are often linked to other activities such as fishing, hunting, and gathering, accentuating the complex web of relationships that exists between humans, animals, plants, Water, and Land. For example, among the Shinnecock, the harvesting of clay was often linked to activities such as clam digging and fishing.

In addition to their connections with other cultural practices, clay harvesting practices also reflect the social and economic structures of the Shinnecock Nation. It was often a communal activity, with families and neighbors working together to gather and process the clay, reflecting the importance of cooperation and mutual support in Shinnecock culture. Shinnecock artist Tohanash Tarrant recounted a story on the importance of Shinnecock pottery, "My grandfather's name Thunderbird came from a vision that he had of his grandfathers in the Shinnecock Hills, and they gave him a shard of pottery. And on that shard of pottery was a Thunderbird. And years later in an archaeological study, they unearth the shard with a Thunderbird, and this Thunderbird became the symbol of our clan. And it also became the vision of starting this powwow and a resurgence of our culture" (Tarrant, 2020). The passage provides a poignant testament to the profound relationality of clay and pottery within the Shinnecock Nation. It showcases the transformative power of ancestral visions and the sacredness of cultural artifacts. The shard of pottery, bearing the Thunderbird symbol, transcends its being a mere object. It becomes a living connection to ancestors, carrying their wisdom and stories through the generations. The discovery of the shard during an archaeological study further underscores the importance of rematriation, the return of cultural knowledge and ways of being to women, family and community. It symbolizes the reclaiming of Indigenous knowledge and practices that were once suppressed or marginalized by colonization. This narrative serves as a powerful reminder that pottery is not solely for practical use but for ancestral memory, cultural resurgence, and the reclamation of Indigenous identity. The use of clay in practical and spiritual activities reflects the importance of cultural heritage and identity, while the social and economic structures that support clay harvesting reflect the importance of community and mutual support in Shinnecock culture.

Techniques and Processes in Clay Harvesting for Soil Health and Community Sustainability in the Shinnecock Nation

The materials and tools used in these practices were shaped by the resources available in the local environment, as well as the cultural and artistic traditions of the community. Clay harvesting among the Shinnecock involved the collection of different types of clay from the local environment. The most commonly used clay was found in areas near the Water, such as creek beds, marshes, and beaches. This clay was collected by hand, using stone, shell, bone, and, later, metal instruments, along with woven baskets and other tools. Other important tools in pottery making included shaping tools, which were used to smooth and shape the vessel, and paddles, which were used to compress the clay and create a more uniform surface. Firing tools, such as fuel for the kiln and firing implements, were also important in the pottery-making process. By using locally sourced materials and traditional tools and techniques, the Shinnecock Nation is able to create pottery that reflects our unique cultural heritage and identity.

Indigenous Peoples have long utilized shells as a tempering material in the preparation of clay for firing. The practice of incorporating shells into clay serves multiple purposes and reflects the deep connection between Indigenous cultures and their natural surroundings. Shells, with their abundance in coastal regions, were readily available and offer unique properties when added to clay. Not only do they enhance the plasticity and workability of the clay, but they also play a crucial role in regulating the firing process. During the firing, the shells act as a tempering agent, reducing the risk of cracking or warping by providing an even distribution of heat. This age-old technique demonstrates the resourcefulness and ingenuity of Indigenous Peoples in utilizing natural materials and our profound understanding of clay and firing dynamics (Maclean, Sutphin, and Bankoff, 2022; Newsome and James, 2019).

Contemporary Challenges and Opportunities for Regenerative Clay Harvesting in the Shinnecock Nation

Contemporary extractive practices in ceramics refer to the modern methods of sourcing and using clay that prioritize efficiency and profitability over sustainability and cultural preservation. These practices often involve the extraction of large quantities of clay from the earth through industrial mining or large-scale excavation, which can have negative environmental impacts such as soil degradation, Water pollution, and loss of biodiversity. In addition, contemporary extractive practices may not necessarily respect the cultural significance of clay and its traditional uses among Indigenous communities. This highlights the need for a shift towards more regenerative and culturally sensitive practices in the ceramics industry.

Indigenous practices for harvesting and using clay are often in stark contrast to this industry. Indigenous communities have a deep understanding of the interconnections between human activities and the natural environment, and our traditional practices are often guided by principles of sustainability and stewardship. In contrast, contemporary extractive practices in ceramics are often driven by profit and short-term gain, and can have significant negative impacts on soil health and the environment.

One of the key differences between both practices is the approach to Land use and management. Indigenous communities take a holistic approach to Land use, recognizing the importance of preserving soil health and promoting biodiversity. Traditional Indigenous practices for harvesting and using clay involve selective harvesting, an approach that ensures that the ecosystem is not overexploited and that the surrounding environment remains healthy and sustainable.

In contrast, the use of heavy machinery and equipment, large-scale mining or excavation of clay can damage soil structure and contribute to soil degradation. The transportation of clay over long distances can contribute to greenhouse gas emissions and air pollution, further impacting the environment. Another important difference between Indigenous practices and contemporary extractive practices is the use of natural materials and methods. Indigenous communities, in contrast to conventional firing methods, often utilize natural materials like wood and seaweed in pit firing to fire clay, a process that not only preserves tradition but also enriches the soil with nutrients and organic matter, fostering healthy soil and biodiversity. Conventional extractive practices often rely on the use of fossil fuels and other non-renewable resources for the mining, processing, transportation, and firing of clay. This can also contribute to greenhouse gas emissions and air pollution and can have significant negative impacts on the surrounding environment, including soil degradation, habitat destruction, and the loss of biodiversity.

A new generation of Shinnecock ceramic practitioners are carving out clay knowledge to pass on to future generations. These Shinnecock artists are scientists in their own right, delving deep into Shinnecock knowledge and practice to secure the health of clay and people for future generations. Shinnecock ceramic ecology scholars such as Courtney M. Leonard pursue interdisciplinary methods to communicate art and science woven through

Indigenous storytelling methodologies. More precisely, their work draws attention to the global climate crisis through the lens of clay and co-constituted relationships with coastal ecosystems. Her work also highlights the unique whaling traditions of the Shinnecock who were whalers and were highly regarded as renowned mariners during the height of the whaling industry in the Atlantic. As part of her practice, Leonard aims to recover Whaling logbooks of prior centuries that are used to reconstitute contemporary stories of climate resilience and devastation and challenge Euro-normative temporal spatialities that would relegate ancestral wayfinding to forgotten or, worse, erased histories. The Shinnecock Nation is not the only Indigenous Nation involved in these processes of rematriation of ceramic ecologies. Shawnee Chief Benjamin J. Barnes says of the Shawnee, "Our citizen-scholars seek to reclaim our artistic birthright and resurrect a dormant ceramic artform" (Barnes and Warren, 2022, p. 2). The processes of ceramic rematriation require spiritual, ceremonial, and cosmological awakenings to breathe life into ways of knowing clay that may have been sleeping for many decades, if not centuries, due to colonization, assimilation or other legal prohibitions limiting Indigenous clay practice and access to ancestral harvesting areas.

Shinnecock ceramic ecologist Courtney M. Leonard frequently collaborates with Indigenous communities, experts, and scientists to create multidisciplinary projects that address ecological challenges. By actively engaging with local communities including her own nation, Shinnecock, she encourages dialogue, knowledge sharing, and the preservation of traditional ecological knowledge. Her collaborative approach helps bridge the gap between art, science, and Indigenous perspectives, fostering a holistic understanding of ceramic ecology.

Recently, there has been a growing recognition of the importance of Indigenous-led research and collaboration in promoting regenerative clay harvesting practices. Courtney M. Leonard's work can be considered foundational to Indigenous ceramic ecology due to her focus on exploring the intersection of ceramics, culture, and the environment. Her artistic practice often addresses issues related to sustainability, ecological awareness, and the interconnectedness of humans with the natural world. Leonard emphasizes the significance of clay as a material deeply rooted in the earth. She explores the historical and cultural connections between clay and the Shinnecock, emphasizing the importance of sustainable clay harvesting practices. The artist began a series of artworks released in 2009 exploring the relationship between mapping, clay, and colonialism entitled CONTACT made of ceramics, canvas, brass, and sinew, which includes CONTACT 1609 and CONTACT 2021. In 2023, Leonard expanded on the CONTACT series to create a large installation of clay thumbprints woven together on canvas providing a map of Paumanok (Long Island) (Fig. 10.2, 10.3, and 10.4).

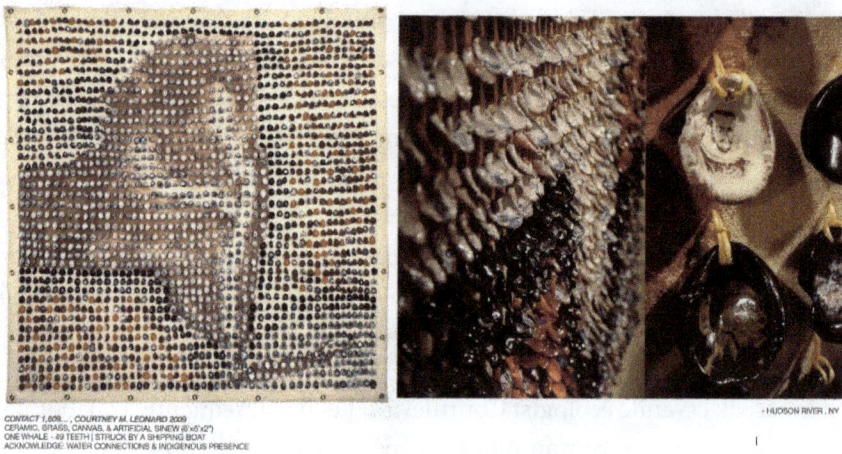

CONTACT 1,609..., COURTNEY M. LEONARD 2009
CERAMIC, BRASS, CANVAS, & ARTIFICIAL SINEW (6'x6'x2")
ONE WHALE - 49 TEETH | STRUCK BY A SHIPPING BOAT
ACKNOWLEDGE: WATER CONNECTIONS & INDIGENOUS PRESENCE

- HUDSON RIVER . NY

Figure 10.2. *Courtney M. Leonard, CONTACT, 2009. Porcelain w/ Fired on Enamel and Red Iron Oxide Transfers, Sewn w/ Artificial Sinew on Cotton Canvas Stained w/ Wedged Red Clay w/ Brass Grommets. 6'x6'x2". Image courtesy of the artist.*

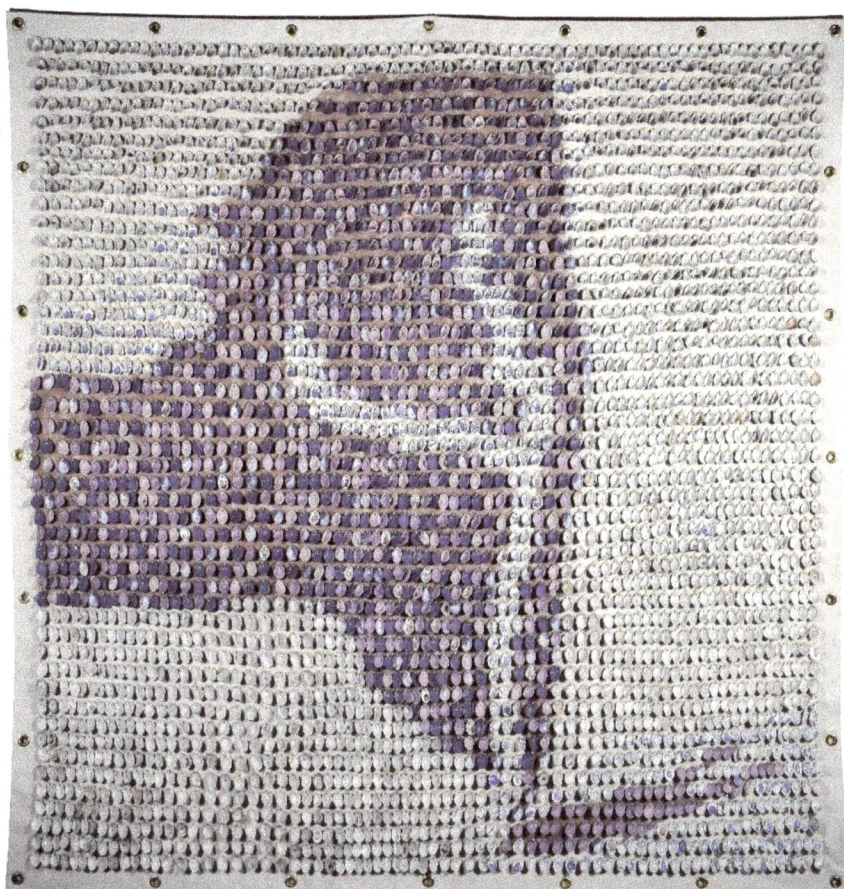

Figure 10.3. *Courtney M. Leonard, CONTACT, 2021. Porcelain w/ Fired on Enamel and Red Iron Oxide Transfers, Sewn w/ Artificial Sinew on Cotton Canvas Stained w/ Wedged Red Clay w/ Brass Grommets. 6'x6'x2". Image courtesy of the artist.*

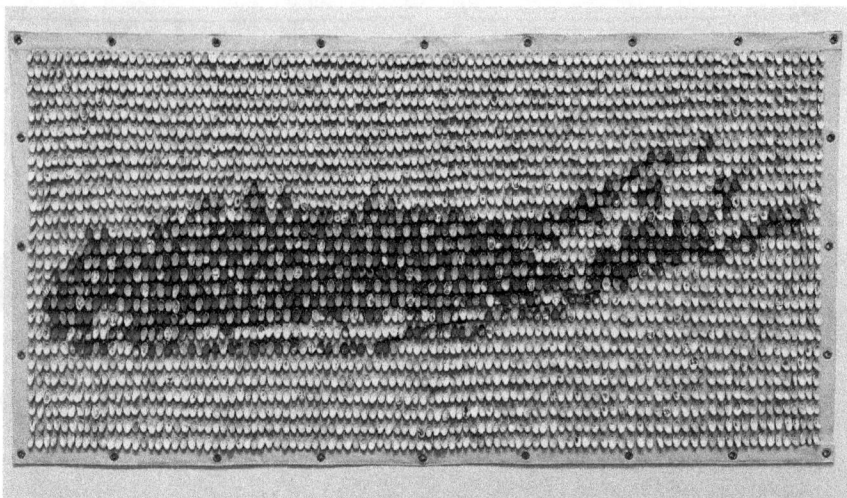

Figure 10.4. *Courtney M. Leonard, CONTACT, 2023. Porcelain w/ Fired on Enamel and Red Iron Oxide Transfers, Gold Luster and Pearlescent Luster Sewn w/ Artificial Sinew on Cotton Canvas Stained w/ Wedged Red Clay w/ Brass Grommets. 6'x6'x2". Image courtesy of the artist.*

By raising awareness about the ecological impact of clay extraction and advocating for responsible sourcing, Leonard promotes a more conscious and ecologically sensitive approach to working with ceramics. Through her artwork, the artist highlights environmental issues and narratives specific to Indigenous communities, often focusing on the impacts of climate change, Land degradation, and the rematriation of cultural practices. By incorporating ecological themes into her ceramic pieces, she brings attention to the relationship between humans and the environment, fostering a deeper understanding and appreciation for ecological concerns.

Barriers to Future Rematriation of Regenerative Clay Harvesting Practices

Regaining control of traditional clay harvesting practices for the Shinnecock Nation is not without its challenges. The impact of colonialism and commercial extractive practices has disrupted traditional land-use practices and degraded soil health, making it difficult for Indigenous communities to reclaim their traditional practices. Moreover, climate change also poses increasing threats to Indigenous access to clay. Long

Island, being a coastal region, is particularly vulnerable to sea level rise (Leonard, 2021). As sea levels rise, coastal erosion can occur, potentially affecting clay deposits located near the shoreline. Increased coastal erosion can lead to the loss or degradation of clay-rich areas along the coast.

Beyond climate change and loss of Land due to erosion and inundation, one of the main challenges facing the Shinnecock Nation is access to non-Shinnecock territory for clay harvesting. The vast majority of Land on Long Island is privately owned, making it difficult for Indigenous Peoples to access the Land we need for harvesting clay. For instance, increased demand for Land for second homes, vacation rentals, or sprawling suburban development leads to the destruction and degradation of clay-rich areas. In addition, the federal process of reclaiming Land is a lengthy and complicated ordeal that often does not bear fruit (Leonard, 2021).

As urban areas expand, they encroach on natural areas, leading to the loss of sensitive and culturally important habitats. The destruction of natural areas also leads to increased erosion, which negatively impacts soil health. The construction of roads, buildings, and other infrastructure also leads to soil compaction, erosion, and loss of topsoil. The use of heavy equipment during construction further degrades soil health. These practices can lead to soil erosion, nutrient depletion, and the loss of soil structure. The impacts of climate change, such as increased temperatures and changes in precipitation patterns, will likely exacerbate negative impacts on soil health. Warmer temperatures will likely increase soil erosion, while changes in precipitation patterns may lead to both soil erosion and nutrient depletion. Pollution from sources such as industrial waste and chemical runoff will further contaminate soils, making them unsuitable for agriculture and other uses.

Other environmental impacts of colonialism include soil degradation, Water pollution, and habitat destruction, which all impact the health and availability of clay deposits. Long Island agricultural practices, such as tillage and the use of chemicals, negatively impact soil health. Clay deposits have been contaminated by pesticide and fertilizer use, which seep into the soil and groundwater. Additionally, the disposal of agricultural waste, such as animal manure and crop residue, also contributes to soil and Water contamination. This contamination has negative impacts on the environment, including the loss of biodiversity and the degradation of soil and Water quality. It also has implications for human health, as contaminated soil and Water impact food and Water sources. Overall, the use of pesticides, fertilizers, and other agricultural practices has had long-lasting and far-reaching impacts on the environment and the Long Island Indigenous Nations and communities that depend on it.

The lack of access to educational opportunities has made it difficult for younger generations to learn about traditional clay harvesting practices. With the forced assimilation of Indigenous children into the colonial education system, many Shinnecock youth are not exposed to traditional practices that in prior decades and centuries would have been passed down from generation to generation. As a result, there is a risk that these practices will fall dormant. These issues highlight the ongoing impacts of colonialism and extractive practices on Indigenous communities and the need for systemic change to support regenerative practices that promote soil health and community sustainability.

Regenerating Healthy Shinnecock Clay Futures

Today, the Shinnecock Nation continues to practice ancestral pottery making, using many of the same materials and tools as our ancestors. By preserving and passing down these traditions, the Shinnecock Nation is able to maintain a connection with our cultural heritage and identity while also promoting the sustainable use of natural resources in our local environment. As we saw, Indigenous clay harvesting practices are important for promoting healthy soils and communities, as well as for maintaining cultural heritage and identity. These practices are deeply rooted in our cultural traditions and have spiritual and ceremonial significance. The use of selective harvesting and natural materials and methods is one way Shinnecocks contribute to promoting and advancing healthy soil and biodiversity in our ancestral territory of Paumanok.

Within this chapter, I have traced the emergence of Indigenous ceramic ecology through the lens of the experiences of the Shinnecock Nation. Indigenous ceramic ecology refers to the intersection of Indigenous knowledge, cultural practices, and ecological principles within the context of ceramics. It encompasses the ways in which Indigenous communities engage with the environment through ceramic practices and sustainable approaches to pottery making. Indigenous ceramic ecology recognizes the deep connection between Indigenous communities and our ancestral Lands and Waters. It involves the transmission of traditional knowledge, techniques, and values related to clay sourcing, processing, and pottery making. This knowledge often incorporates sustainable practices that have been developed and refined over generations. Indigenous ceramic ecology emphasizes the sustainable sourcing of clay and other materials used in pottery making, ensuring the long-term viability of clay deposits. Indigenous ceramic ecology plays a significant role in preserving cultural identity and traditional practices. Clay is an important part of Indigenous cultural heritage, and the ecological principles embedded in pottery-making contribute to the sustainability and continuity of Indigenous communities.

Notes

[1] The capitalization of "Land" and "Water" in this context is intentional and serves to acknowledge the Shinnecock Nation's belief in the living essence and intrinsic relationships of these entities. In many Indigenous knowledge systems, Land and Water are considered living beings, imbued with spirit and agency, integral to the past, present, and future of the Shinnecock people. By recognizing Land and Water as living entities, the significance of their role in the preservation of cultural practices, such as clay rematriation, is underscored. This recognition highlights the deep connection between the Shinnecock and our environment, emphasizing the interdependence and mutual responsibility that forms the foundation for cultural resurgence and environmental stewardship.

References

Barnes, C.B.J. and Warren, S. (2022). *Replanting Cultures: Community-Engaged Scholarship in Indian Country*. State University of New York Press.

Bokuniewicz, H. (1980). Groundwater seepage into Great South Bay, New York. *Estuarine and Coastal Marine Science*, 10(4), pp.437-444.

Landsman, N. (2011). The Worlds of Lion Gardiner, ca. 1599–1663: Crossings and Boundaries: Special Issue Introduction. Early American Studies. *An Interdisciplinary Journal*, 9(2), pp.241-247.

Leonard, K. (2021). WAMPUM Adaptation framework: Eastern coastal Tribal Nations and sea level rise impacts on water security. *Climate and Development*, 13(9), pp.842-851.

Merrill, F.J.H. (1883). On the Geology of Long Island. *Annals of the New York Academy of Sciences*, 3(1), pp.341-364.

Newberry, J.S. (1895). *The flora of the Amboy Clays (No. 26)*. US Government Printing Office.

Newsom, B.D. and James, M. (2019). Beyond Grit and Shell: Evidence of Conifer Needle Temper in Archaeological Ceramics from Maine, *Archaeology of Eastern North America*, 47, pp. 135–155. Available at: https://www.jstor.org/stable/10.2307/26816286.

National Museum of the American Indian (NMAI) (2022). *Pipe*. Available at: https://www.si.edu/object/pipe:NMAI_106073

MacLean, J.S., Sutphin, A. and Bankoff, H.A. (2022). Indigenous Peoples Before the City. *Buried Beneath the City* (pp. 17-50). Columbia University Press.

Shinnecock Indian Nation. (2019). *Shinnecock Indian Nation Climate Vulnerability Assessment and Action Plan*. Available at: https://www.peconicestuary.org/wp-content/uploads/2019/10/Shinnecock-Indian-Nation-Climate-Vulnerability-Assessment-and-Action-Plan.pdf

Smith. M. (2022). A View to a Kiln. *The Long Island History Project*. Available at: https://www.longislandhistoryproject.org/a-view-to-a-kiln/

Tarrant, T. (2020). *Tohanash Tarrant: Shinnecock Portrait Project*. Available at: https://shinnecockportraits.com/tohanash/

The Long-Islander (1839-current). May 23, 1896, Page 1. *NYS Historic Newspapers*. Available at: https://nyshistoricnewspapers.org/lccn/sn83031119/1896-05-23/ed-1/seq-1/

Chapter 11

Ittoqqortoormiit Through the Pinhole: Exscribing the Ethno-Aesthetics of Pia Arke

Morten Søndergaard

In the practice of Pia Arke, "digging earth" refers to a double extractivism: one that is an actual extraction from the territories of Indigenous lands but revisited and retold (and reimagined) as a geopolitical boundary breaking point leading to a possible new position of herself as artist; and another, more personal extractivism, an extraction of herself or an extractivism that is "taking colonialism personal," as Arke writes (Arke, 2010, p. 13).

Her work is founded in the dislocation and the hybridity of her double identity, being of both Greenlandic and Danish descent. She continually questions how to operate as an artist being from Greenland with a Danish father and an Inuit mother, with training in an art academy in Copenhagen. Intensively challenging the technoscientific culture of the West and its role in the construction of a postcolonial identity, she is left within a zone of endless doubts and interpretations, bringing her to a geo-ethnographical brink, searching for that which Edward Said referred to as "beginnings" of understanding postcolonial reality and psychology, the interconnect of lands and persons, geography, and identity. (Said, 1975).

In the process, she grapples with the relationships among many disparate feelings and things, fragments and forgotten places, memories, and people, investigates all kinds of methods—artistic, ethnographic, anthropological, scientific, and works at reremembering them. So, even though the existing artistic modes of expressions and the aesthetics posited by the art world that she encountered at the art academy in Copenhagen in the 1990s provoked the beginning of her personal extractivist project, they soon proved inadequate for continuing it; something was missing. What was missing, I argue, is what the rest of her artistic life is and will be about: the formulation of a transethnic aesthetics by collecting the material and ideas for its foundation into a very personal, paracolonial archive.

In this chapter, I will show how she develops a method of exscribing, in the sense of Jean-Luc Nancy and later refined by Wendy Chun (Nancy,2013; Chun, 2012), this archive of forgotten, untold, and unheard people and memories by reading and writing, showing them, and relating them elsewhere as they are, as she writes, unfinished and delayed: "In my delayed and unfinished settlement of accounts with colonial history I include a lot of stories that family and friends in and outside Scoresbysund, Greenlanders and Danes have opened up for me, a confused cluster of memories attached to and released by these photographs taken in Scoresbysund and spread for the north wind. I make the history of colonialism part of my history in the only way I know by taking it personally" (Arke, 2010, p. 13).

Unwriting and Rewriting

The significance of Arke's extractions is in the distinct methods she develops, as part of her artistic practice, of rereading and rewriting of histories in the light of traces of colonialism.[1] Arke uses a method that draws on the ideas of exscription, which according to Wendy Chun is a process of reremembering through reformatting, or reading by writing elsewhere (something that fits well to Arke, I would argue, even if Chun is writing about digital memory):

Memory as a collective act is evident in our daily experiences with computers, which do make their memory seem—against physics— permanent and ubiquitous. If things remain now it is not because digital memory is more robust than paper; the opposite is true. Silicon degrades far more quickly than paper; websites—even seemingly as solid and permanent as geocities—disappear. Clouds are clouds: mercurial mixtures of vapor, liquid and solid. Despite this, digitization can be considered a form of "saving" because it preserves content through a process of "reformatting": a reading, which is writing elsewhere. All computers read by writing elsewhere, or, to use Nancy's terms, by exscribing: by copying and disseminating. Importantly, exscription has little to nothing to do with meaning, but everything to do with communicating (Chun, 2016, pp. 18–19).

The art project/work *Nuugarsuuk Alias Pinhole Camera Photograph Alias the Point* (1990) refers to two locations: Nuugarsuuk (in Dundas near Thule in Western Greenland) (Fig. 11.1) where Pia Arke went to school; and interreferentially, Ittoqqortoormiit (Scoresbysund in northeastern Greenland), where Pia Arke was born.

Figure 11.1. *Pia Arke: Nuugarsuuk Alias Pinhole Camera Photograph Alias the Point, 1990. Silver gelatin prints on Baryte paper. Credit: The Pia Arke Estate.*

In the 1990s, she revisited both sites and performed an artistic intervention using an oversized camera obscura, which she had constructed at the art academy in Copenhagen and then transported to Greenland. (Kleivan, 2009, p. 251) With this very large camera obscura construction, in fact, the size of the houses where she used to live at the same sites, the mountains and mines surrounding her childhood worlds and the lost structures and cultures, stories about herself, and her heritage are gradually revealed.

Some things, however, are not dug up and remain silent and invisible. Like the Denmark-Greenland relationship or the colonialism contextualizing them. In this way, I claim (and we are still with Chun and exscription), that Arke is activating memory, or, rather, using the pinhole camera, (among a variety of other strategies) to obscure (as it were) what we think we know and remember of the postcolonial life and condition, and replace it with memory as an activation-tool, because, as Chun writes, "Rather than arguing whether memory exists and what is or is not memory, what is most needed is a change of perspective: one that acknowledges that memory is an action, an activation and difference in structure, making perhaps, memory not anything because it is everything" (Chun, 2016, p. 18).

The pinhole pictures are like dark prisms signifying the impossibility of operating an ethno-aesthetics, at the same time posing the question: how, then, should you operate as an artist in this situation? How does an artist with a hybrid background navigate the paracolonial condition surrounding her? To answer such questions, she needs urgently to take all memories out of storage and rewrite and regenerate them, essentially extracting them from their postcolonial condition. "Memory is not storage," as Chun reminds us; "Memory is not something that simply remains: if any memory remains it is because it is constantly regenerated. The conflation of memory and storage is dangerous because it fosters a misleading ethos that forgets the collective care and effort—good and bad—that goes into any memory" (Chun, 2016, p. 18).

During her relatively short life, Arke developed a method where the question and critique of extractivism takes on a double meaning. It is enhanced and refers to the digging of earth as a matter of creating beginnings, which, as Edward Said has pointed out, is not about origin "divine, mythical and privileged" but is "secular, humanly produced, and ceaselessly re-examined" (Said, 1985, p. xiii). Furthermore, this focus on new beginnings is her way of exscribing her own story, essentially accumulating evidence in images, texts, and ideas, again very near to Said's elaboration of beginnings as a poetic and epistemological category, which "include[s] . . . such trends as the critique of domination, the re-examination of suppressed history . . . the cross-disciplinary interest in textuality, the notion of counter-memory and archive, the analysis of [invented] traditions . . . professions, disciplines, and corporations" (Said, 1985, p. xiii). Still, even as this is a rather fitting description of the context and outer layers of Arke's practice, her method includes even more silent layers and personal considerations.

It is one thing to begin deconstructing the culture of colonialization you were born into, another to work toward new and different ways of being and existing as an artist that do not entail old oppressive oppositions or imaginaries. New imaginaries are needed, based on the rewriting and archiving of the silent stories from the Greenlandic-Danish entangled lives around Ittoqqortoormiit in northeastern Greenland.

Ethno-Aesthetics and Beyond

"With a slight pressure on the shutter release, the Greenlanders have a long tradition of 'moving' south with their European progenitors, that is, in depicted form, as images" (Arke, 2010, p. 12).

In her thesis from 1995, Pia Arke analyzed this paradoxical and complex situation of being an artist from Greenland in a modern, Western artworld. The concept and reality of the expectations of an ethno-aesthetics, she demonstrates convincingly, is in itself a European construction which she cannot escape, because "ethno-aesthetics can be analyzed as an event combining ethno-centrism and anthropological humanism at which . . . the only people actually present are the people that are not displayed, i.e. the Europeans." Therefore, she continues, "my ethno-aesthetic issue with ethno-aesthetics constantly runs into the inconceivable fact that the suppressed remains suppressed no matter how conscious I am of its suppression" (Arke, 1995, p. 20).

Arke builds her argument partly on a text by the Italian philosopher Mario Perniola, in which he states that art is neither ontologically related to its place or culture of origin, an idea which he refers to as ethno-aesthetics, nor is it constantly nomadic; instead, he argues, art in a postdisciplinary, generic, aesthetic situation displays simultaneity, which means that it is simultaneously situated in the spaces we visit and live in, and therefore operating in transitory conditions: therefore, Perniola argues, art is interface—to other things than art, or art as medium of the *umwelt*, an environment reflecting and intertwining the surrounding world. Art as simultaneity, according to Perniola, does not rely on the metaphysics of identity, but "presumes reassessment of the in-betweens, of l'entre-deux, of a Zwischen" (Perniola, 1994, p. 6).

Arke concludes that "there is a sense of urgent necessity about our play with the pieces of different worlds" (Arke, 1995, p. 35). This notion I would claim is significant and indicative of the way she approaches postcolonial discourse, and developing her own method, as it is observed by Kuratorisk Aktion in the book Tupilakusaurus: "Arke remains cool and does not try to cut the knot with a heroic stroke of the sword, even though the sword that was called postcolonial studies was on offer from abroad. Instead, she slogged away in one work after the other to loosen that knot and, if possible, arrange it a little more humanely" (Kuratorisk Aktion, 2010).

Arke challenges herself (and us, her audience) to work with colonialism and what she terms as *ethno-aesthetics* as relative concepts (Rogoff, 2012), in the sense that they are as much about giving voice to what has happened as they are about finding and formulating that which she calls a third position—as an artist and as a person, effectively replacing her focus on *beginnings* as opposed to *origins*, to use Said's terms (Said, 1985, pp. xii–xiii).

Models of Thinking in Postcolonial Discourse

According to art critic and writer Irit Rogoff, it is possible to distinguish between three models of thought in postcolonial discourse, defined here as the critical analysis of the relations of colonialism and their ongoing impacts (Rogoff, 2012, p. 329). The question regarding the application and relationality of the concept of postcoloniality is relevant in the case of any inquiry into the processes, politics, cultures, economies, and aesthetics of digging earth across a multiplicity of places and temporalities.

Complex Political Geographies

The first model of thinking is largely based on Rogoff's reading of Edward Said's thought on British Marxist literary critic Raymond Williams. In his text, Said discusses the notion of a situated consciousness and the other, with particular focus on the colonizer and the colonized. He proposes a dual model for analyzing separately the complex political geographies of the colonizer and the colonized, where each is saturated with systems of thought and ideas that cannot be made fully operational or functional due to the question of a possible native vision (Said, 1990). The impossibility of the native Inuit position is embedded in all of Pia Arke's work, and her dialogue with the Swedish intellectual and journalist Stefan Jonsson, for instance, plays on this duality of complex geographies stated by Rogoff here. In Arke's thinking, making the case for another consciousness is an additional strong element; it represents a different and third space of her highly singular hybrid critical consciousness.

Postcolonialism as Ideology

The second model of thought is a largely a Marxist reading of the colonized, the working proletariat, as in opposition to the colonizer as the face of capitalist systems of exploitation. In this model of thought, the focus is on the opposition of a true and a false invasive position (of former colonizers) between which there is no dialogue. There is no sensitivity toward complex or hybrid situations. This is the postcolonial position that Arke seeks to circumnavigate, not because she does not feel the pain from the oppressive iterations of Danish colonial rule or European colonialism in general, which is certainly very present in her life and practice. Rather, she is keen on finding and formulating a cooler way to address this and to allow for the activity of exscribing, rerembering and rewriting memories and people in another place and time than where they existed.

Contact Zones

The third model of thought is concerned with contact zones, defined as "social spaces where disparate cultures meet, clash and grapple with each other" by the anthropologist Mary Louise Pratt (Rogoff, 2012), between those complex geographies and the people navigating them. Furthermore, they often meet in "highly asymmetrical relations of domination and subordination, colonialism, slavery, plantation culture etc." (Pratt, cited in Rogoff, 2012). Instead of periodizing and historicizing a narrative of ethnic groups arriving at certain moments and causing rupture or disaster, creating a sort of rationalization for tyranny or subjectification, the notion of contact zones changes the perspective completely; according to Pratt, there have always been contact between peoples across complex geographies and the traces of these 'zones' of contacts can be seen in the shifts occurring gradually in the ways people behave, think, imagine, and live their biotechnical lives[2]. So, contact with different cultures is not just a painful condition, but something that pushes us around, challenges us, and changes the way people in the 'zones' live, the ideas they live by, and how they express them.

Boundary Situations

Contact zones envisioned as a model of critical postcolonial discourse are certainly detectable in Arke's works and projects. To an extent, it is possible to see her entire oeuvre as one big contact zone and the hybrid singularities as inscribed by a myriad of contacts, and for Arke, first and foremost, on a very personal level, between her Greenlandic mother and Danish father and then between the cultures and geographies of Greenland, the coastal cultures of Søndre Strømfjord and Thule, also the culture of the American base in Thule where she grew up, and additionally Denmark, as modern European culture and polity. The list goes on. Arke's work is deliberately inscribed in the contact zones of science and art, Western (anthropological) research methods, and her own ideas and methods of inquiry into her past, identity, self, and gender. Contact zones, and what they produce, can be seen as boundary situations, to expand on the term *boundary object* coined by Jack Burnham and later elaborated by theoreticians Susan Leigh Star and John Law (Star, 1995). And these boundaries, transduced by Arke, expose a de novo a language of art, a system of art practice that is not limited to traditional media, aesthetics, and genres (in the sense of Howard Becker, 1982, pp. 362-4). As Burnham observed, "A 'sculpture' that physically reacts to its environment is no longer to be regarded as an object. The range of outside factors affecting it, as well as its own radius of action, reach beyond the space it materially occupies. It thus merges with the environment in a relationship that is better understood as a 'system' of interdependent processes. These processes evolve without the viewer's empathy. He/she becomes a witness. A system is not imagined, it is real" (Burnham, 1969, p. 10).

Soil and Stories

Arke's complex method is especially evident in what is arguably one of her most extensively exscriptive works, *Stories from Scoresbysund* (Arke, 2010). In what is essentially an artistic research project published in the format of a book, she unwrites and rewrites her own early story in Ittoqqortoormiit, exscribing a silent part of her own and her family's life from archives of photos, letters, and diaries as well as scientific studies—all of which are located elsewhere, in attics and basements, libraries, and city archives in Denmark and other places in Europe. Stated on the back of the book, documenting the project, is: "In order to gain a historical consciousness of the place and her own origins visual artist and artistic researcher Pia Arke (1958–2007) has tracked down and unearthed all this evidence and out of it created a fascinating collage of pictures and stories" (Kuratorisk Aktion, 2010, p. 56).

Scoresbysund has a peculiar history, as this bay area in the scarcely populated northeast coast of Greenland was colonized in 1924 by the State of Denmark, settling a dispute with Norway. In 1925, Inuit families from other locations one thousand kilometers south of Scoresbysund were moved to this new makeshift frontier location by the Danish colonial authorities, among them Pia Arke's mother.

Stories from Scoresbysund tell those stories that would otherwise have remained silent and unwritten. However, the focus does not appear to be condemnation, but merely reportage about who was there and how they lived. Along with *Soil for Scoresbysund*, it is above all relational, similar to the contact zone relationality in the way Burnham describes it (Burnham, 1969). All the photos, information, letters, and diaries unfold as a network of relations from which an elaborate and refined systemics slowly appears. It is within this systemics that the contours of a paracolonial discourse becomes detectable in Arke's practice. A philosophy emerges from behind the scenes, from the silences and the collection of fragments; it is an epistemological questioning that challenges every construction of meaning that has been inscribed to Greenland and Denmark and to anything interpretable as colonial or of ethnic beginnings.

In *Soil for Scoresbysund* (see Fig. 9.2), we witness a process that stages a multiplicity of contact zones operating in a system of independent relations that merge with the environment on various levels. Soil becomes a medium—the physical medium as well as the vehicle—for the contact zones and the potential activity exscribed. And that is indeed one important message in Arke's work: that the exscription of what was inscribed by other people, other histories, other cultures, in different times, and reading and writing them again in new contexts and other spaces become the soil for future biotechnical lives in Greenland.

Figure 11.2. *Pia Arke, Soil for Scoresbysund (1998). Installation of coffee filters, coffee, and cotton string. Constructed from description in letter to Stefan Jonsson. Photo: Anders Sune Berg. Credit: The Pia Arke Estate.*

That Arke's work points toward beginnings as an epistemological category is clear in this somewhat atypical work. It seems atypical because Arke only rarely worked sculpturally, but I argue that even in this case the sculpture is not the primary meaning-carrying aesthetic form but rather a sort of poetic elaboration of her method: exscription and extraction of a combination of stories and soil exposing the meeting of very different worlds. The packages of soil, which refer to the coffee grounds that the new civilization in Scoresbysund threw out in the trash heaps outside the makeshift frontier houses after the daily coffee drinking are traces of human activity shared by Greenlanders and Danes alike. The packages look like gifts wrapped in cheap paper and bound with sail twine. They hint at being unpacked and opened, to view the beginnings of new meaning they might contain. The knots have a dual function, to keep the soil in place inside the package and to invite their opening. They designate a boundary and implicate a network of possible relations – between the various boundaries of people living in extreme nature and acting as liminal spaces of Arke's personal memories. Unopened, however, they remain poetically silent and secret and compost for later excavations: "It is three people's consumption of coffee over four months. When I was in Greenland, I was going to throw out the coffee grounds, but I was told that I shouldn't do that, as they wanted it to become compost. They so much wanted to have some more soil. After a sufficient length of time coffee becomes earth. Coffee is imported; it is something people drink all over the world." (Arke, 2010).

Situating Knowledge of the World at the Edge of the World

So, even though her work can be seen as an exploration of her own identity and unsettled and perturbed practice beyond an ethno-aesthetics (Arke, 1995), her lasting contribution to the topic of this volume is her elaborative and interventionist extraction of the concept of digging the earth as a relative concept, oscillating between the metaphoric, mythic, and poetic and the concrete, material, and scientific. By inscribing her own body elsewhere, in between geographic, and geological placements, she positions her postcolonial practice and person at the beginning(s) of a de novo art world thereby activating an aesthetic system positioning her as a boundary subject in a paracolonial condition.

To better understand this paracolonial meaning construction in her work, a little context is needed. The very notion of an art world and the aesthetic system that dominates it, and controls the inclusion in and exclusion from it, could be envisioned as a genealogy of aesthetic paradigms that circulate in opposition and exchange with each other (Søndergaard, 2020). It is possible to observe the existence of five such aesthetic paradigms that have dominated art worlds since the eighteenth century and were part of the cultural backdrop to colonialism.

Indeed, the term *aesthetics* was first coined in the eighteenth century by the German art critic Baumgarten, partly to describe the difference between European and Indigenous art. The first paradigm was not formulated or activated as an exclusive European discourse, however, until Baumgarten's term was refined and epistemologically combined with the idea of subjectivity by the German philosopher Immanuel Kant. He sought to describe the faculties of knowledge and judgment in what he perceived as the shocking aftermath of the French revolution and to define the principles of human sense making in an increasingly complex and modern world. Aesthetics thus becomes a matter of sense perception and how the world becomes available for the subject by processes of immanent meaning-creation. *A priori* and *pleasure* are two very different immanent instances of sense making, or rather: the first creates a sense of meaning, the other a sense of beauty. This aesthetic paradigm of subjective sense perception was the dominant paradigm of the art world until the beginning of the twentieth century when both the art world and premodern cultures were destroyed by world wars; the enormous transitions and transformations that occurred has resulted in what is known as modern culture and society.[3] This sense-perception paradigm is still active and has circulated in and out of the art worlds of the twentieth and twenty-first centuries alongside later aesthetic paradigms.

After World War I, a different art world and aesthetic system grew from the ruins of the old world. It was partially inspired by the Russian and failed German revolutions of 1917 and 1919 and the ideological model of

Marxist thought. But it was also inspired by new intellectual, philosophical, and scientific ideas and was generally poised toward experimenting with the forms and formats of art. Art under this paradigm is seen as being culturally constructed, mainly by the ruling classes and the capitalist system. There is a history of this construction which can be rewritten. The idea that humans are estranged from nature (and the nonhuman world) and that a "dissociation of sensibilities" has taken place (Austin, 1962) is a central driving force behind this paradigm. What art hides and transcends rather than what is displayed and can be perceived becomes essential. Indeed, "The Construction of Aesthetics" is the title of the doctoral dissertation (about the Danish philosopher Søren Kierkegaard) by one of the main philosophical architects of this second paradigm, Theodor Adorno (Adorno; Hullot-Kentor,1989). In the 1930s, this paradigm and the older, prevailing one clashed in an ugly ideological war between the far left and far right, which reverberated globally and effectively drew a map of the beginnings of a postcolonial era where the dominant aesthetic paradigms were increasingly structured from.

In the aftermath of the Second World War, culminating around 1968, a third aesthetic paradigm emerges. Driven by cultural, social, and political circumstances as well as new scientific, philosophical, and intellectual ideas, this third paradigm critiqued the focus on perception and sense making, and aesthetics was seen as constructed by structures of ingrained cultural patterns of understanding. These genealogies and their discourses take the very idea of aesthetics in a completely new direction (partly driven by ideas from structuralism and, later, poststructuralism), and for the first time, a colonial critique and a postcolonial discourse emerged alongside civil rights movements, feminism, and a general skepticism toward any authority. In this paradigm, aesthetic meaning is constructed by its reliance on (often hidden) structures. The circumstances and dynamics involved in the third paradigm are not unlike that which activated the previous paradigms: great changes, great expectations, and great fears. This third paradigm was still dominant around the time Pia Arke began to work at the art academy in Copenhagen. And because of the proliferative environment surrounding the theory department led by the Danish philosopher Carsten Madsen, who invited prominent philosophers of this paradigm (Jean Francois Lyotard, Paul Virillio, and Mario Perniola among others) to serve as guest professors, Arke was trained in this way of thinking. Aesthetics and art were about contextualization, including of artists themselves, always entailing a critique of the conceptual and political discourses surrounding art as an activity carried out by a person entangled by memory and bias.

In the early 1990s, however, there were other dynamics at play. The fall of the Soviet Empire, for example, instantly triggered a series of postcolonial conflicts in the former Soviet colonies or territories of influence in Eastern Europe, Africa, the Middle East, and Central and South America. And new discourses on science, technology, and the media affected ideas about art and aesthetics. Two other paradigms (the fourth and fifth) were generated from these conditions and as reactions to the earlier paradigms. The fourth paradigm sees technology as means for our transcendence—that technology determines culture and humanity. And since the relation between humans and the world is technocultural, the argument is, we need new tools and strategies to understand and navigate life, or we risk that technology will dominate (Kittler, 1990; Stiegler 1998).

The fifth aesthetic paradigm draws on social constructivism and actor network theory and from there takes a fresh look at the constructions we live by, based largely, but not entirely on the metaphors of information theory and new social theory. From these ideas, network and relational aesthetics appear, playing on the metaphors of media and computer theory, including cybernetics. This fifth paradigm of aesthetic meaning, based on network thinking and a critique of the very humanistic, earlier paradigms, introduces a slightly more scientific turn for the way art is iterated. Instead of seeing art as expressed primarily and formally with the intention of creating an artwork, art is seen as an activity reacting to other activities. It is argued that rather than approaching art as a 'phenomenon' it should be approached from a techno-anthropological perspective allowing for inquiries into the materialities of art practices and the 'cultures' of media and computing (Serres, 1980; Star, 1995; Latour, 2007).

It is possible to see Arke's practice as placed between the third, fourth, and fifth aesthetic paradigms, and thus between three differing art worlds. The fourth paradigm, however, has not (yet) produced its own art world in the sense that, unlike the earlier paradigms, it has not dominated a globally expanded art world and ways of understanding art. Rather, it is a paradigm which is active in subcultural segments of experimental and culturally critical layers of activities. The genealogy of aesthetic paradigms that are still circulating in the greater theater of our global postcolonial world adds a new and sad layer to the extractivism conducted by Pia Arke, since they are paradigms of exclusion based on ideas structured by European mindsets and paradigms of aesthetic domination. In between those paradigms Arke carved out a space for herself to practice and explore the personal.

Walrus Bay

The shady and slightly obscured rocky icy landscape of Walrus Bay sets the scene for the extraction of Arke's ethno-aesthetic past and lost context; the "natural" technology of pinhole photography captures this site as impossibly transparent and painfully obliterated, spectacular and dark. The photographs signify the complexity of her own situation, torn between her first home and a Western tradition (that she also cherishes) which she encountered at the art academy in Copenhagen.[4]

Figures 11.3-a, b, c. *Pia Arke, Walrus Bay 1944 and 2001. 2001–3. Three Color photographs off the ground where the US weather station in Walrus Bay near Scoresbysund/Ittoqqortoormiit used to be located. In white frame with pencil annotations. Credits: The Pia Arke Estate and National Museum of Photography, Copenhagen, Denmark.*

The investigation in the series is about her self, her identity, but even more, it is about art as a relative concept in need of being extracted. She conducted stunning and impressive revisions of the aesthetic paradigms structuring the art world (in the sense of Howard Becker); and in the process she developed pioneering methods of artistic research in an almost generative sense, looping science and art practices into each other in unexpected ways.

My exscription of Pia Arke's own unwriting and extractivist practice draws partially on an exhibition I cocurated in 1998, Legender / Legends.[5] For this exhibition, Arke created the work *Tupilakusaurus*, the first in a series of works that took her to the heart of silences, the unheard and unwritten of that which I term her paracolonial position (Anonymous, 2012). In this way, the inquiry of this chapter is also an unauthor(ized) auto-ethnographic digging into the necessary premises and (im)possibilities of unwriting the (hi)story of the Denmark–Greenland relationship.

Notes

[1] For an excellent survey of the history and cultures of Greenland, see https://www.cs.mcgill.ca/~rwest/wikispeedia/wpcd/wp/h/History_of_Greenla nd.htm

[2] I use this term as designating liminal spaces as "a technical" form or formal technics, indeed a general mediality that is constitutive of "... the human as a "biotechnical" form of life. Media, then, functions as a critical concept in something like the way that the Freudian unconscious, Marxian modes of production, and Derrida's concept of writing have done in their respective domains." (Michell and Hansen, 2010)

[3] See Matai Calinescu, *Five Faces of Modernity*. (Calinescu, 1987).

[4] Philosophers Jean Francois Lyotard, Mario Perniola, and Paul Virillio among others were associate professors at the Royal Danish Art Academy in the 1990s and had significant influence on the level of conceptual reworking of postcolonial artistic practices.

[5] The exhibition was prepared by my colleague, Tine Seligman. As a freshly appointed curator at the Museum of Contemporary Art in Roskilde, Denmark, I was merely lucky to step in and be part of the exhibition construction process.

References

Adorno, T. W. and Hullot-Kentor, R. (1989.) *Kierkegaard: construction of the aesthetic*. Minneapolis: University of Minnesota Press.

Kuratorisk Aktion (2010). *Tupilakosaurus: An incomplete(able) Survey of Pia Arke's Artistic Work and Research*. Kobenhavn: Kuratorisk Aktion.

Arke, P. (1995). Etnoæstetik. *Ark* (17).

Arke, P. (2003). *Scoresbysundhistorier: fotografier, kolonisering og kortlægning*. Kobenhavn: Borgen.

Arke, P. (2010). *Stories from Scoresbysund: photographs, colonisation and mapping Scoresbysundhistorier : fotografier, kolonisering og kortlægning*. (2ⁿᵈ Ed.). Kobenhavn: Pia Arke Selskabet and Kuratorisk Aktion.

Austin, A. (1962). T. S. Eliot's Theory of Dissociation. *College English*, 23 (4), pp: 309-312.

Becker, H.S. (1982). *Art Worlds*. Berkeley: University of California Press.

Burnham, J. (1969). The Aesthetics of Intelligent Systems. *Reel-to-Reel collection. Solomon R. Guggenheim Museum Archives*. Available at: https://www.guggenheim.org/audio/track/the-aesthetics-of-intelligent-systems-by-j-w-burnham-1969

Calinescu, M. (1987). *Five faces of modernity: modernism, avant-garde, decadence, kitsch, postmodernism*. Durham: Duke University Press, 1987.

Chun, W. H.K. (2016). Ubiquitous Memory: I Do Not Remember, We Do Not Forget. In Ekman U., Bolter

J.D., Diaz L., Søndergaard M. and Engberg M. (Eds.), *Ubiquitous Computing: Complexity and Culture*, Routledge, 2016, pp. 161–74.

Kittler, F. A. (1990). *Discourse networks 1800/1900*. Stanford: Stanford University Press.

Kleivan, I. (2012). Pia Arke's Pinhole Camera Photographs, 2009. *Tupilakosaurus: An Incomplete(able) Survey of Pia Arke's Artistic Work and Research*. Kobenhavn: Kuratorisk Aktion.

Latour, B. (2007). *Reassembling the Social. An Introduction to Actor-Network-Theory*. Oxford: Oxford University Press.

Mitchell, W. J. T. and Hansen, Mark B.N. (2010). *Critical Terms for Media Studies*. Chicago: The University of Chicago Press.

Nancy, J.L. (2013). The Exscription. *Alea: Estudos Neolatinos*. 15 (2), pp: 312-320. Available at: 10.1590/S1517-106X2013000200004.

Perniola, M. (1994). Art as Neutral Mutant. *45th Venice Biennale: Cardinal Points of Art Theoretical Essays*. Venice: Marsilio.

Said, E. W. (1985). *Beginnings: Intention and Method*. New York: Columbia University Press.

Said, E. W. (1990). Narrative, Geography and Interpretation. *New Left Review*, I(180), March-April 1990.

Star, S. L. (1995). *The cultures of Computing*. Oxford: Blackwell Publisher.

Rogoff, I. (2012). Is the Postcolonial a Relative Concept? *Tupilakosaurus: An Incomplete(able) Survey of Pia Arke's Artistic Work and Research*. Kobenhavn: Kuratorisk Aktion.

Serres, M. (1980). *Le parasite*. Paris: Grasset.

Stiegler, B. (1998). *Technics and Time, 1: The Fault of Epimetheus*. Stanford: Stanford University Press.

Søndergaard, M. (2020). Sound as Evidence: Paradigms of Aesthetic Approximation in an Age of Geopolitical Crisis. *Leonardo Music Journal*, 30(84), p. 84. Available at: https://doi.org/10.1162/lmj_a_01096

Chapter 12

DesertArtLAB: Ecologies of Resistance

Rachel A. Zimmerman

Rebranding the Desert[1]

Amid brightly colored multimedia displays, DesertArtLAB's contribution to the *Mahlzeit/The Meal* exhibition in Vienna in 2022-2023 was a white table with simple white place settings, each plate topped with a small cactus on a mound of soil. This display, titled *Desertification Dinner,* presents a jarring contrast to European culinary practices. Indeed, sourcing suitable cacti in Vienna was challenging, as such plants are considered an exotic curiosity in much of Europe. In arid regions of the Americas, cacti have served as essential food sources for millennia. They are integral to human histories and mythologies, such as the nopal cactus that determined the location of Tenochtitlan, now Mexico City, and is featured on the Mexican flag. *Desertification Dinner* raises questions about how European and Euro-American consumers value certain plants, often those associated with Europe, and devalue others. While cactus remains a food source for Indigenous peoples in North America who have resisted the eradication of their foodways, the Euro-American culture that dominates the United States has excluded cactus from the mainstream diet. Within this cultural context, presenting cacti on plates with an accompanying text that reads "salad for a hotter, drier future" is an act of ecological resistance.

As historian Erik Altenbernd writes, the desert was the type of landscape that nineteenth-century white Americans "loathed most and understood least" (Altenbernd, 2016, p. 4). In his 1823 map, explorer Stephen H. Long labeled a large swath of land east of the Rocky Mountains "Great American Desert" (Long, 1823). Today, the simplest definition of a desert is a region with no more than 10 inches of annual precipitation although other definitions also consider evaporation. Long's "Great American Desert" receives 10 to 25 inches of annual rain, while true deserts are found further west. Long and other explorers often applied the term desert to landscapes they perceived as *deserted*—largely devoid of life (Altenbernd, 2016, p. 9-13). Although Indigenous peoples thrived on the local flora and fauna, Long did not recognize the cultural and economic value of a landscape so unlike the East Coast of the United States.

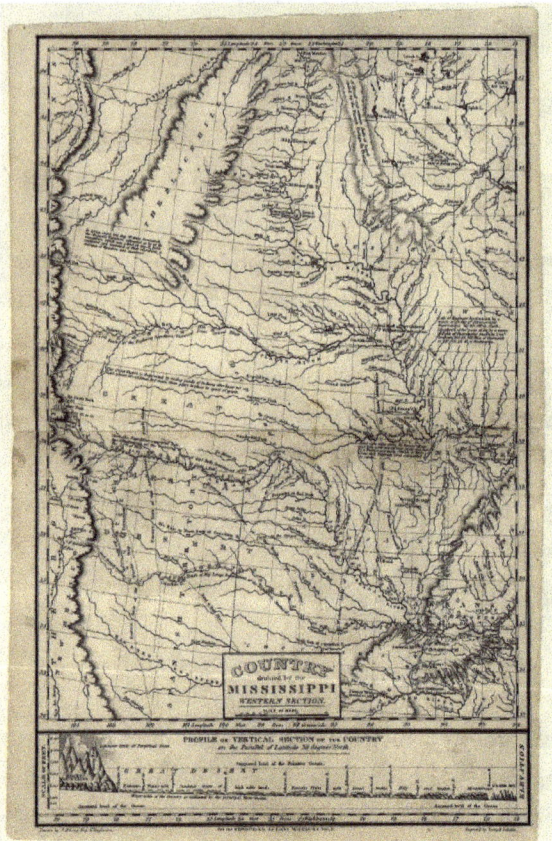

Figure. 12.1. *Stephen Harriman Long, Country Drained by the Mississippi, Philadelphia: Young & Delleker, 1823. Library of Congress, Geography and Map Division.*

Long's map intersects the region with the words "Spanish Territory" and reads "the Great Desert is frequented by roving bands of Indians who have no fixed places of residence but roam from place to place in quest of game," strengthening the perception of the land as foreign, unfit for cultivation, and wild (Long, 1823). Such stereotypes continue to shape perceptions of the desert in the United States. City codes, homeowner associations, and representations in television, film, and books enforce the ideal of green lawns of imported grasses, bushes, and trees that derives from European landscape painting and garden design (Jenkins, 1994, pp. 10-13, 16).

With gallery displays, planting projects, and community events, DesertArtLAB works to dispel the perception of the desert as hostile and barren. Professor Matt Garcia (Chicano) and museum professional April Bojorquez (Chicana/Rarámuri descent) founded DesertArtLAB in 2010 and have had several collaborators, and repeated support from the Institute of American Indian Arts in Santa Fe. Unlike other examples of Land Art that reshape the land to suit the artist's goals, Bojorquez and Garcia highlight the aesthetics of the desert. For them, the desert is the site of their family histories, community, and foodways. The end-products of DesertArtLAB artworks range from digital images to cooked meals, incorporating desert plants, Indigenous knowledge of arid ecologies, and acts of resistance against destructive and oppressive agricultural and ecological practices.

For generations, the artists' families have lived in the dry climates of Arizona and Southern Colorado. Indigenous nations inhabited these lands for thousands of years before the region was claimed as part of the viceroyalty of New Spain, later renamed Mexico, and finally incorporated into the United States in 1848. Beginning in the 1960s, the Chicano Movement encouraged some people who had been labeled Mexican American to recognize and embrace their Indigenous ancestry (Blackwell, 2017, p. 102-4). Affirming the Indigeneity of Chicanx history, customs, and recipes is a form of resistance against past and present methods of cultural genocide that attempt to eradicate Indigenous languages, religions, and ways of life. As is common among Chicanos, Bojorquez and Garcia do not hold tribal affiliation but the recipes that their families have prepared and eaten for generations connect them to their Indigenous ancestry.

Ecological Interventions

Although in recent years DesertArtLAB's work has been included in numerous exhibitions, the artists collective is founded on outdoor planting projects, community involvement, and adaptation to the local environment. The collective formed in Phoenix, Arizona, April Bojorquez' family home. With an average of eight inches of rain per year, Phoenix is not only a desert but also a particularly hot one. Summer temperatures above 100 degrees Fahrenheit are typical with temperatures nearing 120 degrees becoming ever more common. Yet, Phoenix is advertised as a leading destination for golf, with immense amounts of water diverted to replace the desert with non-native grasses. Additionally, a survey from 1998 determined that 42.6% of central Phoenix consisted of vacant lots (Pagano and Bowman, 2000, p. 4). In 2010, Matt Garcia and April Bojorquez began a series of performances during which they planted nopal cacti in neglected patches of soil around town. These resilient plants needed no further watering or maintenance and could grow into nourishing sources of food, providing the city with life and encouraging an appreciation for the desert ecology.

The following year, Bojorquez and Garcia planned a community mass-planting event in an empty lot. The project, called *Parque Libertad/Fields to Freedom*, united about one hundred people to plant dozens of nopal pads that had been rescued from an uprooted and discarded cactus. The event was a joyous celebration of community, food, and the desert but ended in disappointment. Once the cacti had been planted, the landowner arrived and ordered the plants to be removed. In a culture that despises the desert, empty dirt is preferred to living cacti.

In 2013, DesertArtLAB continued their community outreach efforts with the *Mobile Eco-Studio*, a tricycle equipped with storage and a work surface. Bojorquez and Garcia rode the tricycle around neighborhoods and taught families how to make and plant desert seed mixes. Bojorquez, Garcia, and the residents of the neighborhoods collectively shared their recipes for preparing and eating nopales. Participants were given seeds and nopales to plant in vacant lots to further nurture, with community activism rather than official permission, the nourishing desert ecology.

Continuing the focus on the cultural, ecological, and culinary importance of dryland plants, Garcia and Bojorquez designed *Field Site*, a long-term ecological installation that incorporates numerous stages, collaborations, and harvests. *Field Site* will contribute to the creation of a *Desertification Cookbook* over the course of several decades. The landlord's interference at the *Parque Libertad* demonstrated the need to purchase a plot of land to enable long-term growth. In 2016, the artists used funds from a Creative Capital Award to acquire the most abused piece of land that they could find in Pueblo, Colorado, Matt Garcia's family hometown. Pueblo's land has been subjected to severe extraction and pollution. Several smelters operated from 1878 to 1921 and the steel mill has been in use since 1881 (Fry, 2000). The pollution from one silver and lead smelter has led the Environmental Protection Agency to designate portions of the town a Superfund site (Balkin, 2021, p. 236). The *Field Site* is located on a patch of compacted empty dirt that had been used as a road between a paved parking lot and residences, a couple miles upriver from the Superfund site.

As of 2022, *Field Site* has undergone two phases of preparation and planting. *Field Site Phase One* in 2016 involved documenting the condition of the site, surveying the area, and collecting any artifacts, such as discarded car parts and other trash. Planting began with Indigenous community members performing a blessing. During the first phase, the team planted 134 chollas, a cactus that can withstand high heat and severe cold with little water, can grow to ten feet in height, and gifts edible flower buds. Over the years, the chollas' exceptionally strong roots have carved into the soil making water, oxygen, and nutrients more available and enabling other plants to grow. In 2021, the second phase began with Ballet Folklorico performers planting yucca. As the site matures, DesertArtLAB will continue to enhance biodiversity and food value with additional plant varieties and will begin to harvest and prepare meals from the site's plants.

Knowledge

One of DesertArtLAB's early projects begun in 2011 was titled *LAND knowledge*. Gathering and sharing ecological knowledge is central to DesertArtLAB's purpose. While at the Museu Nacional de Antropología in

Mexico City, Bojorquez and Garcia encountered the text of a Nahuatl huhuetlatolli that resonated with their work:

> *Ten cuidado de las cosas de la tierra.*
> *Haz algo, corta la leña, labra la tierra, planta nopales, planta magueyes.*
> *Tendrás qué comer, qué beber, qué vestir.*
> *Con eso estarás en pie, serás verdadero, con eso andarás.*
> *Con eso se hablará de ti, se te alabará, con eso te darás a conocer.*
> *-Huehuetlatolli*
>
> (Act! Take care of the things of the earth.
> Do something, work the land, plant nopal, plant maguey.
> With that, you will have something to eat, to drink, to wear.
> With that, you will stand, be true. With that, you will walk.
> With that, they will speak of you, praise you. With that,
> you will be known.
> -Huehuetlatolli)[2]

The huehuetlatolli are orations that guided ethical behavior among the Nahua, that Spaniards recorded in the sixteenth century. This example affirms the long-standing reciprocal relationship with the land, with people caring for desert plants that, in turn, provide food, beverages, and fibers for making textiles. Honoring ecological knowledge that has guided Indigenous peoples for centuries is part of DesertArtLAB's mission. Indigenous peoples home to drylands have built bodies of knowledge that allow growing, eating, and living suited to arid environments. As Dr. Margaret Kovach (Cree) states in regard to Indigenous knowledge, "Place links present with past and our personal self with kinship groups. What we know flows through us from the 'echo of generations,' and our knowledges cannot be universalized because they arise from our experience with our places" (Kovach, 2009, p. 61). The *Field Site* highlights the importance of place-based knowledge. Bojorquez and Garcia were repeatedly told that they would not be able to grow on the abused and compacted site without purchasing gardening soil and providing supplemental irrigation. Instead, the artists trusted the relationship between the soil and native plants that Indigenous peoples have observed, remembered, and shared for millennia. While the team visits and maintains the site - with regular pruning and weeding - they do not provide water, fertilizer, or soil amendments. Every cholla has survived with soil that was originally on the site and rainfall which has ranged from 15.99 inches in 2017 to 5.33 inches in 2020 (National Atmospheric and Oceanic Administration, 2018; n.d.). Healing the land in ethical and sustainable ways relies on the knowledge of people who have lived in communion with drylands for generations.

In 2021 Garcia and Bojorquez began creating a series of cyanotypes that engages with the history of knowledge production and the interconnection between people, plants, and the environment (Figure 12.5). Cyanotype is an early photographic process that creates blue prints. As an artform, cyanotypes are most associated with British woman Anna Atkins' nineteenth-century botanical studies. She placed plant specimens on photographic paper to create intricate silhouettes, labeled them with scientific names, and bound the images together into books. Garcia and Bojorquez adopted this process to make photographs of the desert plants that they grew. Unlike Atkins' delicate small-scale photographs, DesertArtLAB's are large-scale prints. In some, April Bojorquez includes herself in the images, thereby melding the human body and plants. The photographs are created outdoors with the sun exposing the treated paper.

Food knowledge is key to April Bojorquez' and Matt Garcia's relationship to the desert as a source of nourishment and connection to their cultural heritage. Ethnobotanist Dr. Enrique Salmon (Rarámuri) describes the significance of sharing food recipes to Indigenous "knowledge reproduction" and thus to the persistence of Indigenous cultures (Salmon, 2012, p. 7). The *Mobile Eco-Studio* incorporated this sharing of recipes as Bojorquez and Garcia taught their methods for preparing nopal to community members and, in exchange, learned recipes and their accompanying stories from participants. The in-progress *Desertification Cookbook* gathers and shares recipes for preparing desert plants as well as investigating methods for living in communion with the environment.

DesertArtLAB demonstrates the adaptability of Indigenous knowledge and its importance for the present and future. One vestige of the white supremacist origins of anthropology that continues to shape public perceptions is the treatment of Indigenous peoples as perpetually ancient. Museum dioramas depict Indigenous peoples as static objects of study, and portrayals on television and the internet focus heavily on historic representations (Hill, 2000; Leavitt et al., 2015, p. 40-43). These portrayals have stereotyped American Indigenous cultures as lacking complex technologies relevant to the present. However, in the realm of food, numerous Indigenous technologies facilitate life in arid climates such as processing corn with lime to increase the absorption of nutrients; methods for transforming the toxic and irritating agave into food and a source of fiber; and using *ollas* to create an automatic watering system that responds to soil moisture (Salmon, 2012, p. 140; Buchanan, 2022; Bainbridge, 2015, pp. 17-20).

Such denial of Indigenous technology stems from the white supremacist patriarchal capitalism that privileges certain kinds of technology (Beliso-De Jesus and Pierre, 2020). For centuries, Europeans justified colonization, in part, with narratives of their own technological superiority, denigrating non-Western technologies and those associated with women's work as simple and evidence of limited intellect (Adas, 2014). After World War II, the term high technology was increasingly applied to electronic and computer technologies. The United States, at this time strengthening its reputation as a dominating world power, maintained associations between high technologies and the advancement of civilization (Adas, 2014, p. 538-561). Many high-tech industries bring great profits for companies in the so-called Global North at the expense of ecosystems elsewhere. For example, electric cars and solar panels are heralded as solutions to the fossil fuel crisis, yet they rely on lithium. The immense water consumption required for lithium mining is endangering Indigenous communities in Argentina, Chile, and Bolivia (Jerez et al., 2021). High-tech industries effect climate change, the destruction of lands, and the extinction of plants and animals, yet are promoted as more advanced than Indigenous technologies (Nobrega and Varon, 2021).

Resistance

DesertArtLAB's various ecological projects use knowledge and labor to create mutually beneficial systems as the huehuetlatolli advises. This contrasts with the Western approach to landscape conservation that aims to preserve a "pristine wilderness" (Fletcher, et al. 2021). The creation of National Parks, for instance, involved expelling Indigenous peoples and limiting human interference with the land (Hernandez, 2022, p. 79). As environmental scientist Dr. Jessica Hernandez (Binnizá & Maya Ch'orti') declares: "Conservation is a Western construct that was created as a result of settlers overexploiting Indigenous lands, natural resources, and depleting entire ecosystems" (Hernandez, 2022, p. 72). One consequence of this model of land management are the wildfires in the west of the United States that are increasing in frequency and severity. Environmental scientist Dr. Melinda Adams (N'dee San Carlos Apache) has researched how the Indigenous practice of cultural burns alleviate wildfire conditions (Adams, 2020).

The DesertArtLAB collective provides the *Field Site* with regular care to help a healthy ecosystem to form. In 2019, cochineal beetles threatened to slowly kill the chollas. Spraying water to remove the insects has allowed the chollas to recover from the damage, continue to grow, and improve the soil. Over time, the site will create more food for insects, people, and other animals. This process involves closely observing the land, plants, and insects, learning when intervention is beneficial, and working with the desert ecology rather than modifying and eventually degrading the landscape to suit the growth of food plants adapted to humid climates.

Embracing Indigenous knowledge is an act of resistance against white supremacist epistemologies that denigrate non-Western research methods. On November 30, 2022, the White House formally recognized Indigenous knowledge as a legitimate source for informing government policies and procedures. The guidance published on this occasion defines Indigenous knowledge as "based on evidence acquired through direct contact with the environment and long-term experiences, as well as extensive observations, lessons, and skills passed from generation to generation," and states that

Indigenous knowledge does not need to be validated by Western scientific studies (Executive Office of the President, 2022, pp. 3-4). This guidance is significant as white scientists and policymakers have frequently dismissed Indigenous knowledge as unreliable pseudoscience.

DesertArtLAB's working methods not only reframe the value of knowledge but also incorporate anti-authoritarian reclamation of land. The White House's guidance on Indigenous knowledge merely encourages federal agencies to engage in "consultation and collaboration" with Indigenous nations (Executive Office of the President, 2022, p. 2). On the other hand, Land Back, an international Indigenous movement, argues that Indigenous peoples should have sovereignty over the land, making decisions independently rather than as consultants (Pieratos, Manning, and Tilsen, 2021). DesertArtLAB has often worked within this framework of reclaiming land for their work and their community. They began their first planting projects in neglected spaces without approval from landowners or government officials. Their projects also treat land as communal as is common in many Indigenous cultures (Hernandez, 2022, p. 72).

Moreover, the *Mobile Eco-Studio* confronted a racially oppressive political atmosphere. In 2010, Arizona passed SB1070, an anti-immigration law that effectively allowed racial profiling against Central Americans of Indigenous descent (Lopez, 2011). As a result, people at risk of being targeted had to remain in their neighborhoods to avoid police checkpoints. Garcia and Bojorquez, thus, brought their work to oppressed neighborhoods with the *Mobile Eco-Studio*. They considered their work an act of food justice, affirming immigrants' foodways, and encouraging them to grow food and shape their surroundings in the face of racial oppression. As Salmon writes, "eating a bowl of posole made from locally grown corn, lamb, and chili is equal to going to the state legislature to demand fair water rights. This is because when one either grows one's own ingredients for the posole or barters or purchases the foods from someone who did, one is supporting a resilient process aimed at sovereignty" (Salmon, 2012, p. 148).

Field Site engages with anti-capitalist and liberatory practices. At the beginning of the project, Bojorquez, Garcia, and three recipients of a paid summer fellowship provided a service to the community: free cactus pruning. They planted the resulting cholla cuttings at the *Field Site* so each could grow into a new plant. In this way, the artists resist the commodification of food, relying only on resources that are freely available: rain, the local soil, and existing cacti. The second stage of planting reclaimed a resource necessary to life yet commodified under capitalism: water. Colorado has some of the strictest water rights laws in the United States. Until 2016, collecting rainwater on ones' own property was forbidden and the 2016 law permits merely 110 gallons in total storage volume (Colorado General Assembly, 2016). River water is likewise strictly controlled. During the planting, a Ballet Folklorico group gathered water in small pots from the nearby Arkansas river to provide the newly planted yucca with water. Gathering river water was an act of civil disobedience against the treatment of water as private property.

A Hotter, Drier Future

DesertArtLAB enacts these ecological approaches with an eye toward the future. About 41.2 percent of the earth's land is currently considered dryland, and climate change will likely cause drylands to expand and droughts to become more severe (United Nations, n.d.; Means, 2021). DesertArtLAB responds to the expansion of drylands by encouraging ecologically oriented ways of living with the desert. Working with, rather than against, the land and climate presents a stark contrast to the ways that real estate developers have treated dry areas of the United States. Hot desert regions of the Southwest experienced a population boom once the heat could be vanquished with air conditioning and the same kind of housing that is typical of cooler regions could be built quickly and cheaply (Badger and Blinder, 2017). The energy consumed with air conditioning in places like Phoenix exacerbates climate change, leading to increasingly extreme temperatures and unsustainable demands on resources.

DesertArtLAB challenges the cultural norms that favor golf courses over cacti. With longstanding family ties to Arizona and Southern Colorado, Bojorquez and Garcia experience these regions as places of life rather than as barren and inhospitable. As artists, they recognize the power of aesthetics to affect human behavior. Standards of beauty shape our perceptions of how landscapes should look, are enforced by city codes, algorithms, and other societal forces, and can cause people to harm themselves as well as other people, animals, and plants. Like the countless dangerous methods of shaping bodies to the newest definitions of beauty, frequently watering, mowing, fertilizing, and applying herbicide to perennial grasses in arid regions prioritizes the aesthetic of a uniform green lawn over the health of ecosystems which includes humans. In response, DesertArtLAB advocates for seeing the colors and textures of arid lands as beautiful.

Notes

[1] I wish to thank Matt Garcia and April Bojorquez for inviting me to write about their work and for tirelessly answering my questions.

[2] The English translation is by the artists and is sometimes included in their gallery displays. This huehuetlatolli was recorded by the friar Andres de Olmos in the sixteenth century (León-Portilla, 2006, pp. 236-7).

References

Adams, M. M. (2020). Digital Storytelling through Fire: The Revitalization of Northern California Cultural Burns. *Public Scholars UC Davis*. Available at: http://publicscholars.ucdavis.edu/melinda-adams/

Altenbernd, E. L. (2016). *Great American Desert: Arid Lands, Federal Exploration, and the Construction of a Continental United States*. Ph.D. dissertation, University of California, Irvine. Available at: https://escholarship.org/uc/item/1151n78k

Badger, E., and Blinder, A. (2017). How Air-Conditioning Conquered America (Even the Pacific Northwest), *The New York Times*, August 14. Available at: https://www.nytimes.com/2017/08/04/upshot/the-all-conquering-air-conditioner.html

Bainbridge, D. A. (2015). *Gardening with Less Water: Low-Tech, Low-Cost Techniques; Use up to 90% Less Water in Your Garden*. North Adams, MA: Storey Publishing.

Balkin, A. (2021). Visualizing Atmospheric Politics. In: Demos, T. J., Scott,, E.E and Banerjee, S. (Eds.). *The Routledge Companion to Contemporary Art, Visual Culture, and Climate Change*, pp. 230-241. New York: Routledge.

Blackwell, M. (2017). Indigeneity. In: Vargas, D. R., Mirabal, N. R. and Fountain-Stokes, L. L. (Eds.). *Keywords for Latina/o Studies*, pp. 100-105. New York: NYU Press.

Beliso-De Jesús, A. M. and Pierre, J. (2020). Special Section: Anthropology of White Supremacy. *American Anthropologist* 122 (1), pp. 65-75.

Buchanan, A. (2022). *Agave: A Plant with an Intoxicating History*. Washington, DC: Dumbarton Oaks. Available at: https://lab.plant-humanities.org/Agave/

Colorado General Assembly. (2016). *HB16-1005: Residential Precipitation Collection*. Available at: https://leg.colorado.gov/bills/hb16-1005

Executive Office of the President, Office of Science and Technology Policy, and the Council on Environmental Quality. (2022). *Guidance for Federal Departments and Agencies on Indigenous Knowledge*. Available at: https://www.whitehouse.gov/wp-content/uploads/2022/12/OSTP-CEQ-IK-Guidance.pdf

Fletcher, M., Hamilton, R., Dressler, W. H., and Palmer, L. (2021). Indigenous knowledge and the shackles of wilderness. *Proceedings of the National Academy of Sciences*, September 27, 118(40). Available at: https://www.pnas.org/doi/full/10.1073/pnas.2022218118

Fry, E. (2000). *Smelters of Pueblo*. Pueblo, CO: Pueblo County Historical Society.

Hernandez, J. (2022). *Fresh Banana Leaves: Healing Indigenous Landscapes Through Indigenous Science*. Berkeley: North Atlantic Books.

Hill, R. W. (2000). The Museum Indian: Still Frozen in Time and Mind. *Museum News* 79(3), pp. 40-74.

Jenkins, V. S. (1994). *The Lawn: A History of an American Obsession*. Washington: Smithsonian Books.

Jerez, B., Garcés, I., and Torres, R. (2021). Lithium extractivism and water injustices in the Salar de Atacama, Chile: The colonial shadow of green electromobility. *Political Geography*, 87, 102382. Available at: https://doi.org/10.1016/j.polgeo.2021.102382

Kovach, M. (2009). *Indigenous Methodologies: Characteristics, Conversations, and Contexts*. Toronto: University of Toronto Press.

Leavitt, P. A., Covarrubias, R., Perez, Y. A., and Fryberg, S. A. (2015). "Frozen in Time": The Impact of Native American Media Representations on Identity and Self-Understanding. *Journal of Social Issues*, 71(1), pp. 39-53.

León-Portilla, M. (2006). *La Filosofía Náhuatl: Estudiada en sus Fuentes*. Mexico: Universidad Nacional Autónoma de México.

Long, S. H. (1823). *Country drained by the Mississippi*. Philadelphia: Young & Delleker. Retrieved from the Library of Congress, https://www.loc.gov/item/2007630425/

Lopez, T. (2011). Left Back: The Impact of SB 1070 on Arizona's youth. *James E. Rogers College of Law*. Tucson, AZ: The University of Arizona. Available at: https://law.arizona.edu/sites/default/files/Left_Back%20Report.pdf

Means, T. (2023). Climate Change and Droughts? What's the Connection? *Yale Climate Connections*. Available at: https://yaleclimateconnections.org/2021/08/climate-change-and-droughts-whats-the-connection/

National Atmospheric and Oceanic Administration. (2018). *Annual Summary 2017*. Pueblo: National Weather Service. Available at: https://www.weather.gov/pub/climatePUB2017

National Atmospheric and Oceanic Administration. (n. d.). *Annual Summary 2020*. Pueblo: National Weather Service. Available at: https://www.weather.gov/pub/climatePUB2020

Nobrega, C., and Varon, J. (2021). *Big tech goes green(washing): Feminist lenses to unveil new tools in the master's houses.* Global Information Society Watch 2020: Technology, the Environment and a Sustainable World: Responses from the Global South: Free University of Berlin. Available at: https://giswatch.org/node/6254

Pagano, M. A., and Bowman, A. O. (2000). *Vacant Land in Cities: An Urban Resource*. Washington, DC: Brookings Institution, Center on Urban and Metropolitan Policy. Available at: https://www.brookings.edu/wp-content/uploads/2016/06/paganofinal.pdf

Pieratos, N. A., Manning, S. S., and Tilsen, N. (2021). Land Back: A meta narrative to help indigenous people show up as movement leaders. *Leadership* 17(1): 47-61. Available at: https://doi.org/10.1177/1742715020976204

Salmon, E. (2012). *Eating the Landscape: American Indian Stories of Food, Identity, and Resilience*. Tucson: University of Arizona Press.

United Nations (n.d.). *Global Value*. Available at: https://www.un.org/en/events/desertification_decade/value.shtml

United States Environmental Protection Agency. (n. d.). *Superfund Site: Colorado Smelter, Pueblo, CO*. Available at: https://cumulis.epa.gov/supercpad/SiteProfiles/index.cfm?fuseaction=second.cleanup&id=0802700.

Ecologies of Resistance: DesertArtLAB

Figure 12.2
Desertification Dinner, 2022
Courtesy of the artist, © DesertArtLAB.
Exhibition view "The Meal,", Dom Museum Wien, 2022.
Photo: DMW / Kerstin Schuetz-Mueller

Figure 12.3
Field Site, 2016 - present
Courtesy of the artist, © DesertArtLAB

Figure 12.4
Plant Cactus, Grow Cactus, 2017
Archival Maps (Courtesy of Pueblo City-County Library Special Collections)
Laser cut paper,18 x 22"
Image: John Joe at IAIA Museum of Contemporary Native Arts.

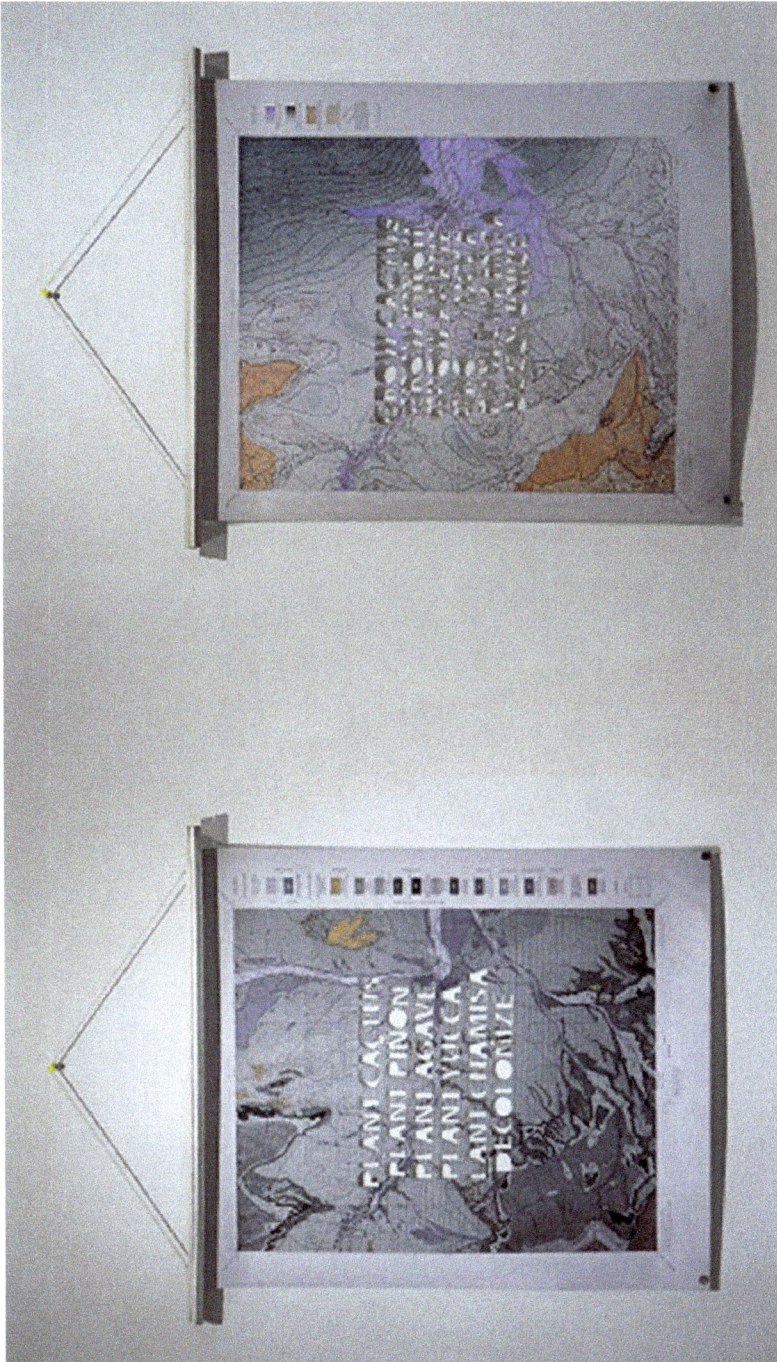

Figure 12.5
Cyanotype, 2021 - present
4 x 5' and 4 x 6'
ink on paper
Courtesy of the artist, © DesertArtLAB

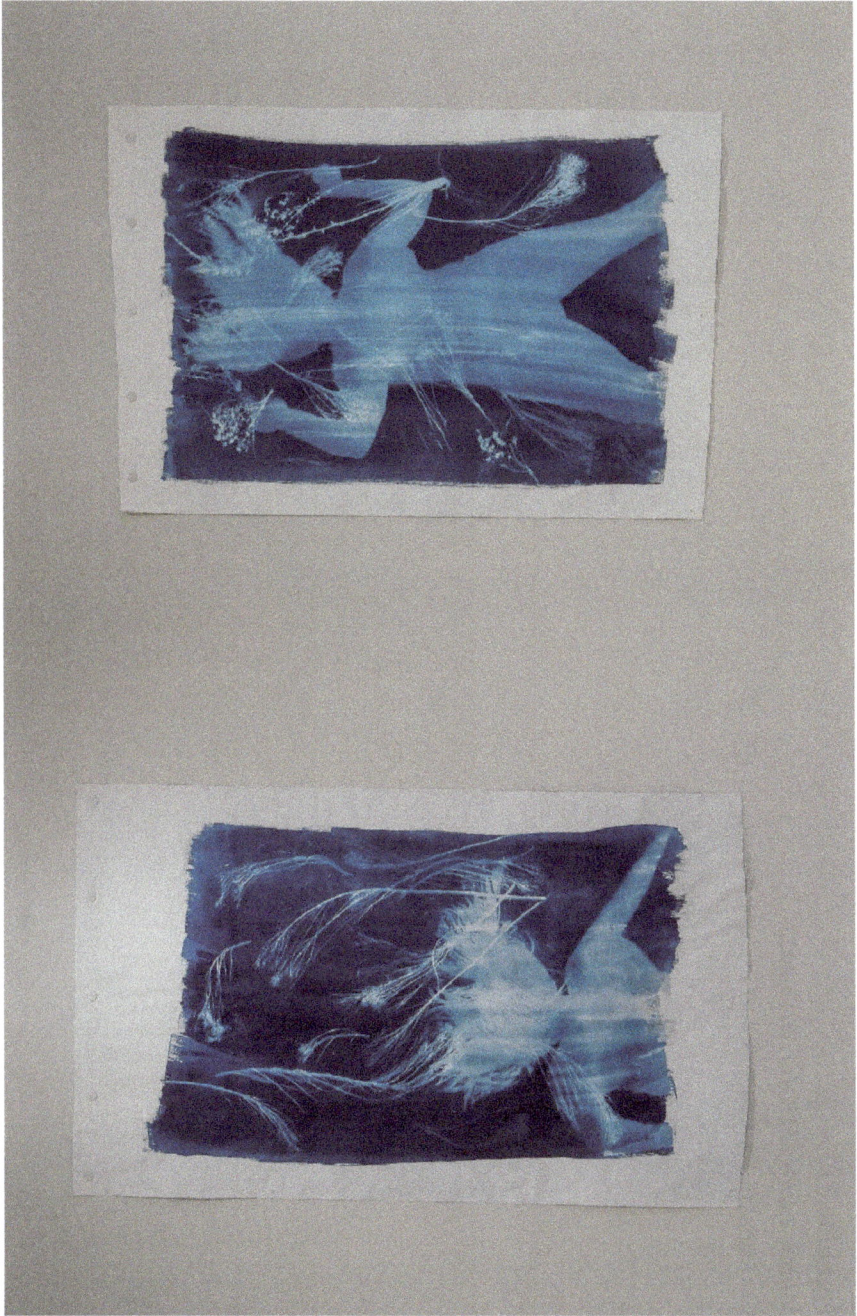

Figure 12.6
Mobile ECO-STUDIO, 2013-2022
Courtesy of the artist, © DesertArtLAB

ACT! TAKE CARE OF THE THINGS OF THE EARTH. DO SOMETHING, WORK THE LAND, PLANT NOPAL. PLANT MAGUEY. WITH THAT, YOU WILL HAVE SOMETHING TO EAT, TO DRINK, TO WEAR. WITH THAT, YOU WILL STAND, BE TRUE. WITH THAT, YOU WILL WALK. WITH THAT, THEY WILL SPEAK OF YOU, PRAISE YOU. WITH THAT, YOU WILL BE KNOWN.
- Huehuetlatolli/Words of the Elders

Chapter 13
Humid Telepathy

Juan Pablo Pacheco Bejarano

When sailors return to land after months at sea, they often have the impression that the ground ripples in the same way as the ocean surface does. What generates this feeling of imbalance? Most scientists believe that when at sea for a prolonged period of time, the brain generates a dynamic modulation in opposition to the movement of water. The result is a feeling of stability that makes us forget the movement of the ship after a few days.[1] On returning to port, sailors feel as if the dry land has turned into water, reducing the abyss between the terrestrial and the maritime worlds. This humid experience of space, neither completely wet nor entirely dry, opens the portal that I propose to explore here.

When I started writing this text I felt an unbearable vertigo every time I sat down in front of my computer. Such was the feeling of imbalance that I thought I had become completely disenchanted with writing; words appeared as a frozen cage for my intuitions. This disenchantment took an unexpected turn when, after a conversation with the Colombian poet Ramona de Jesús, I understood the vertigo I felt in front of the screen to be the same as that of sailors returning to port. Somehow, my land was beginning to moisten. So, I thought: if writing is like getting on a ship for months, what seas have I been on and what ports have I returned to?

What follows is a journey through three ports where my imagination has docked over the past few months, in search of expanding my ability to sense radically different ways of understanding technology. These ports are shaken by the vertigo I have felt in returning to writing from the seas of intuition, as I seek to involve my readers in this vertiginous and humid experience. Slowly I begin to see that my research around water is nothing more than a response to the call of rivers, lakes, and seas in which I have bathed, who ask me to moisten my thoughts and my words. Reason, dry and sterile, is a failed project. Intuition, humid and fertile, will give voice to the deep and distant waters that desire to be heard.

Thanks to the reading group of my first draft with Zenaida Osorio, Bárbara Santos, Maytik Avirama, Mariana Murcia, and Lucas Jaramillo, I understood that telepathy allows me to think precisely about the way we involve ourselves with what is distant from us. Unlike the institutional discourses around telecommunication technologies, which omit any reference to the infrastructures that allow the flow of messages, I am interested in thinking about our involvement with those things that are in-between—that which mediates the transmission of a message. Faced with the socioenvironmental collapse in which we are immersed, what does a reflection on technology, water, and telepathy contribute to our ability to imagine other realities? I want to propose humid telepathy as a tool to feel-think what is in-between, to sustain the life of what is close to us and what is distant from us at the same time. I hope that this humid perspective on the digital revolution will allow us to imagine technology beyond the systems of extraction that currently sustain it. The following ports seek to ripple the earth of our imagination and explore technologies that embrace the interdependence between the digital world and the material world, between territory and information.

Between the Visible and the Invisible

At the first port, I was told that writing about telepathy inevitably invokes the world of the invisible. When I mention the word telepathy to those close to me, the most common image that springs to mind is that of one person transmitting precise information to another—usually a word or a number—without the need for language. Telepathy seems to involve direct communication that does not depend on the translation processes of common mediation. This understanding of telepathy is closely related to a long-standing ambition of modern technoscience: to eliminate the apparent limitations of distance and increase the speed of transmission. The common use of the preposition *tele*, which accompanies almost all inventions of electronic communication, seeks to emphasize this distance eliminated by technical devices: *tele*vision, *tele*matics, *tele*phone, *tele*graph, and recently, due to the pandemic of COVID-19, *tele*shopping, *tele*working, *tele*medicine or *tele*-education. Rather than reminding us of the time and space that our information traverses as it travels through the network, the common use of the preposition highlights our ability to eliminate distance in seeing, speaking, writing, shopping, and working. The modern imagination understands telepathy—the ability to sense the other and the world at a distance—from the elimination of that distance between the bodies sending and receiving a message. In this way, the extraction and transmission of information is maximized and the illusion that capital can be detached from the material world, floating from cloud to cloud, from abstraction to abstraction, is strengthened.

However, the electronic revolution relies on a series of material infrastructures interconnected at a planetary level. Between millions of devices, thousands of data centers, and hundreds of undersea cables, the digital cloud is more like a colossal kraken on the seabed extending its tentacles over land. The term *infrastructure* is commonly understood as a stable, material background on which social relationships develop. Yet, infrastructure is a relational concept that allows us to sense the hidden processes that we usually take for granted. Even more, it refers to the network of symbolic and material systems that give meaning to reality (Star and Bowker, 2006). Although modern infrastructures aspire to the homogenization of procedures by concealing their presence from our visual and narrative spaces, technology and its tangible systems are an active agent of the material and symbolic relations of the spaces and times they occupy. In this sense, reflecting on technological infrastructures allows us to see the relations that occur in the spaces in between, and to invoke the kraken-machine on which the digital revolution depends.

Having invoked our monster ally, my attention turns to the maritime dimension of the internet infrastructure, composed of signal repeaters, submerged data centers, and undersea cables that have been installed since 1858. From its inception, this infrastructure has been closely related to the history of colonialism and capitalism in three main ways. First, the sea routes followed by these cables emerged from the trade routes that mobilize the global market, which in turn originated in the routes traced by the colonial fleets during the first decades of European imperial expansion (Pacheco Bejarano, 2021).[2] The transatlantic flow of consumer goods and digital data that sustains contemporary capitalism converges in the transoceanic infrastructures inaugurated by Columbus. Since then, the purpose of these infrastructures has been to maximize the speed of extraction and commercialization of material and informational goods. Between ships and transoceanic cables, the ancient metropolis is still able to control its territories at a distance, exercising a telepathy aimed at the control, extraction, and infinite reproduction of capital (Pacheco Bejarano, 2020).[3]

Second, digital devices, cables, and data centers are built with minerals such as gold, quartz, cobalt, silicon, lithium, or tin, typically extracted from mines in the Global South. When these infrastructures become waste, they are exported back to these same locations in order to keep the streets of the North pristine, polluting the ecosystems of the South and the communities that care for them. Third, the Silicon Valley monsters emerging from the depths of the digital revolution design interfaces for ever faster transactions—video calls with no delays, passwords stored in DNA—and ever more immersive virtual worlds. In this way, computer multinationals manage to extract data at breakneck speed and sell it to the highest bidder, fueling the unbridled growth of contemporary information capitalism (Zuboff, 2019). Little by little, a global minority withdraws into the digital world while the rest live between the spectacle of screens and the ferocious extractive practices that destroy their territories.

In this essay I will not engage in an extensive discussion of each of these three domains of extraction that define digital infrastructures, as I have discussed them elsewhere. However, I want to highlight how the principle of extraction is deeply linked to the separation between information and infrastructure, between the measurable and the ineffable, between the invisible and the visible. Already in 1977, the Chilean artist Juan Downey commented that "wars against humanity and nature (i.e., the violent extraction of the fruits of the earth) have constituted both the raison d'être and the incentive for the urgent development of technology. Misapplied technology generates apparent wealth, but in the process creates dissonance in the interaction between humanity and nature" (Downey, 2013c, p. 252). More than forty years later, the war described by Downey becomes a planetary crisis, due to a technological imagination that understands land and oceans as resources and as passive containers for the flow of goods. In this scenario, how else might we understand telepathy, that powerful ability to sense ourselves at a distance?

In his phenomenological study on the gestures of modernity, Vilém Flusser suggests that the prefix *tele* implies our approach to distance, an intersubjective dimension of relating (Flusser, 2014). The distant always implies the near, the possibility of understanding what concerns and affects us, what binds us with that which we cannot see and do not experience up close. Thus, when we know something that is different or distant, we are, at the same time, knowing what is familiar and close. Flusser's words ripple the lands of this first port, inviting me to think of telepathy from a more complex dimension of communication, beyond the suppression of distance and time between the nodes of a network. In a more voluminous and profound sense, telepathy refers to our relationship with otherness and the capacity we have to engage in relationships with what is different and distant to us, both in the material and immaterial dimensions. Unlike the obsession of modern technoscience with optimizing the transfer of discernible information, the telepathy that interests me refers to the capacity of bodies to sense at a distance; in other words, by *pathos* as the center of communication. I will return later to the question of the body in relation to telepathic communication.

Although the electronic telepathy that we perform with our computers seems to eliminate what is in-between, distance always implies a spatial and temporal interstice that we cross, both at high and low speeds, between us and that which interests us, that which affects us. In the space between the visible and the invisible, conceiving telepathy from a relational and material perspective allows us to imagine beyond the division between subjects, objects, and their material environments, a logic that characterizes both coloniality and modernity (Mignolo, 2011). Technology constantly dialogues with the biological and geological substrates it traverses and which, to a large extent, sustain it. Energy and matter are inseparable. In order to reformulate our relationship with technology it is imperative to make this shift toward the connections between digitality and materiality. In other words, my interest lies in understanding what sustains the possibility of telepathy, from servers and undersea cables, to the water, minerals, and plants that allow us to store and care for our shared memories. Human beings participate in this web of energetic transference, that surpasses us and at the same time embraces us, from where knowledge is generated and transmitted on a planetary level. I return to the sea from this invisible port, crossing my screen like a water surface, a membrane to other worlds that reflect our images distorted by the waves of their rippling movement.

Telepathic Perspectivism

Some years ago, I dreamt that the ocean flooded all the continents, exceeding the most pessimistic predictions about climate change. Faced with such a scenario, we humans had found a way to separate our consciousness from our bodies. Otherwise, it would have been impossible to survive submerged in the ocean's salty water without corroding the epidermal membrane that protects our internal tissues. Those who had managed to survive the catastrophe inhabited millions of cables that snaked through underwater ruins, installed just before the dissolution of bodies. The fiber optic cables that had sustained the internet for decades had multiplied exponentially to prevent the human species from disappearing, allowing it to prolong its existence in the form of electrical impulses.

In the subtle space between my dream and reality, I dock at a port full of mirrors where the seas and the air of the earth are heating up. Although modernity has always dreamt of the possibility of separating mind and body, a paradigm that permeated my own dream, the acute ecological crisis constantly reminds us that this Cartesian desire is a nightmare unanchored from the reality we inhabit. Separated from the material world, the narratives that construct our relationship with digital technology are insufficient and limited. From the minerals that make up microprocessors—such as lithium, brass, quartz, gold, silica, or cobalt—to the infrastructures that enable the storage and distribution of information—such as data centers or fiber optic cables—, digital technology is closely related to the mineral world. The exploration of this geological dimension of media raises questions about the deep spatial and temporal roots of contemporary media, as well as about their possible futures (Parikka, 2015).

Ecological and material awareness of digital technology has been a fundamental axis of the work of artists and thinkers in Latin America since the mid-twentieth century, guided by the encounter of multiple technological ontologies—or cosmotechnics (Hui, 2021). Inspired by the techno-spiritual experiments of the 1960s and mobilized by a desire to return to Latin America in the wake of the coup d'état to the socialist government of Salvador Allende,[4] Juan Downey made a series of trips between New York and Chile, Mexico, Guatemala, Peru, Bolivia, and Venezuela between 1973 and 1977. The goal of his journeys was to create an inter-American audiovisual network, exchanging videotaped accounts from the hand of Indigenous communities throughout Latin America, projecting in each community what had been filmed along the way. Through this project, called *Video Trans Americas,* Downey envisioned a system of videoconferencing among Amerindian peoples two decades before the popularization of the internet as a globalized means of communication. In sum, this long-term exchange sought to generate a decentralized network of ontological feedback among the Indigenous peoples of the continent, where each community could see itself observing itself and observing others.

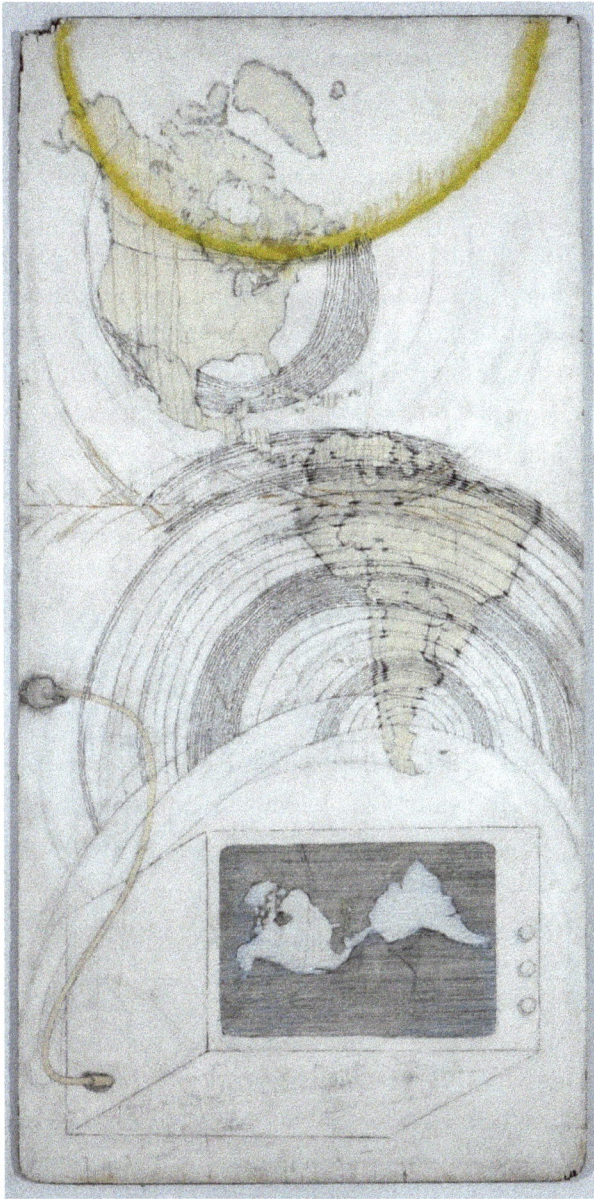

Figure 13.1. *Video Trans Americas, a collage of oil, acrylic, and graphite on wood, made by Juan Downey between 1973 and 1976 while he was travelling throughout Latin America. As a result of this journey he would produce a 14-channel installation under the same name. Indefinite deposit of the Reina Sofía Museum Foundation, 2019.*

The last trip of this project took Downey to the Venezuelan Amazon, where he lived with Yanomami communities. There he shared the closed-circuit television system, which interested the Yanomami because it did not freeze time through a capture, but allowed them to see a space from another perspective at the same time. Downey commented that closed-circuit television allows one "to observe oneself by observing, it increases the concentration of the mind" (2013b, p. 271). His anthropological exercise went beyond the act of looking at others, prompting the viewer "to see oneself by seeing how those others saw him" (Guagnini, 2008, p. 100) thus creating a sort of technological perspectivism. Framed within the systemic thinking of cybernetics, where bodies and territory make up a network of relationships that sustains life on the planet, Downey's approach to technology and its possibilities is marked by an interest in the dynamics that maintain the balance between human beings and their environment, based on communication as the central axis (González, 2013). In this sense, communication is nothing more than the encounter with otherness, a process of simultaneous reflection and self-reflection between beings and their environments.

Three decades later, artist Bárbara Santos began a series of projects with Indigenous communities of the Colombian Amazon. From 2005 to the present day, the collaborative research developed by Santos seeks to build bridges between the ancestral technologies of Amerindian peoples and the digital technologies of modernity. In interviews conducted for her book *Healing as Technology*, several Amazonian knowers speak of their use of feathers, plants, quartz, and other minerals to connect with their homes from afar as well as to care for the multispecies relationships within their territories (Santos, 2019) The electronic devices that fuel the digital revolution, such as sound and image recorders as well as computers, make use of similar minerals to enable communication between places separated by space and time: "minerals in both ancestral and Western knowledge are equally used as systems of recording and memory" (Santos, 2019, p. 83). Although there are multiple differences between the two technological forms, starting with the fact that modern technology depends on the violent extraction of minerals, the comparative and relational research developed by Santos expands what we commonly understand as technology. In this dialogue between two seemingly irreconcilable realities, technology appears as a way of transmuting energy and matter, where care emerges as a central axis of technological development. Considering technology from the perspective of healing allows us to understand our relationship with what is in-between, a relational telepathy of the visible and invisible environments we inhabit.

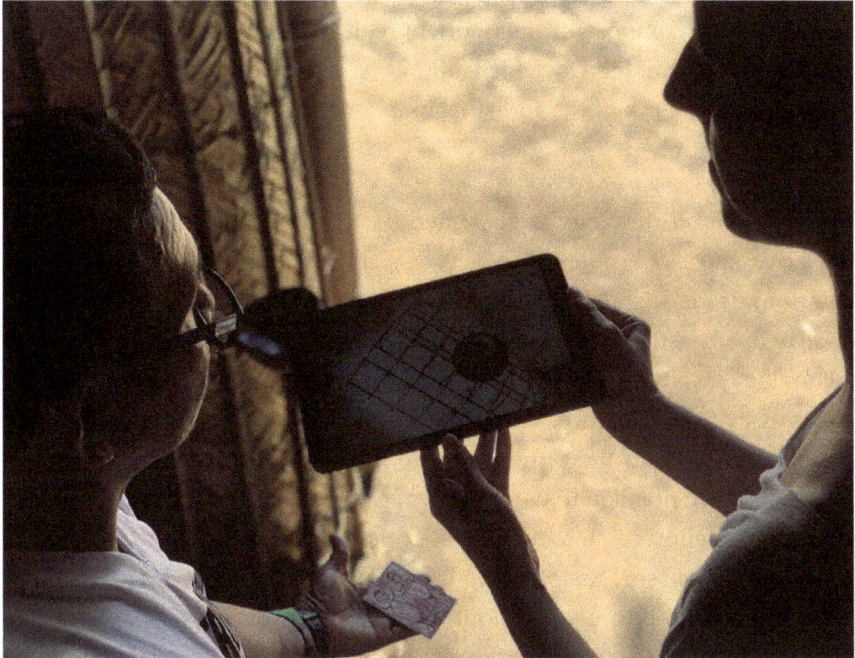

Figure 13.2. *Bárbara Santos (quiasma.co) in the process of cocreating a set of visualizations of ancestral knowledge using augmented reality with the communities of the Pirá Paraná River, as part of her ongoing transdisciplinary project El cuenco de cera (the bowl of wax).*

From his travels and artistic collaborations with Indigenous communities, Downey comments that "in some human beings, brain waves are in symbiosis with natural phenomena: communication with others and with the environment is total" (Downey, 2013c, p. 253). The totality of this type of communication lies beyond the transmission of clear and concise information through language, and is closer to a kind of telepathy that connects through affection and intuition. The traditional knowers of the Amazon care for and administer their territories from *malocas*, houses that enable them to connect to networks of knowledge from within and beyond their geographic location. As Santos says, "sacred places from ancestral knowledge are a network of places, close to what we can understand as a network between nodes of internet servers, a rhizome of dynamic knowledge based on stories of origin, layers of geological knowledge, systemic ecology intrinsically linked to culture and medicine" (Santos, 2019, p. 90). Spiritual navigation, assisted by plants, minerals, and sacred architectures, allows knowers to travel through the multiple bodies and structures that compose this network of reality. The telepathy at the core of Amazonian technologies, which allows knowers to feel and think the ecological networks of the forest from a distance, enables the care of life-affirming relations.

Just as digital technology is sustained by the flow of light and electrical impulses through cables and servers, the technology of Amerindian communities is sustained by the flow of energy through life itself. In the end, both infrastructures are two sides of the same coin, even if digital technology insists on separating its total symbiosis with the material world. In his essay "Architecture, Video and Telepathy," Downey comments that video, thought, and the universe share the same electromagnetic nature, and that "if we manage to enter the wave correctly . . . we can conceive of telepathy, teleportation, and even teleroticism: libidos acting at a distance, collective tantric sex, fusion of lights" (Downey, 2013a, p. 265). Although there is a techno-optimistic echo in Downey's words, characteristic of the expectations tied to electronic developments in the 1970s, the telepathy he speaks of rejects the productive control that dominates the digital technologies of our era. Similar to Santos, Downey imagined the possibility of decolonizing technology as an apparatus of extraction and domination, reconfiguring its power to generate global telepathic networks to reconcile and care for difference. Although Santos's work has a more extended and collaborative dimension than that developed by Downey, both artists expand my understanding of technology through an awareness of the deep relationship between the visible and the invisible worlds. To decolonize technology is precisely to recognize this interstitial materiality, the relationship of our bodies and territories to the technologies that connect us.

The distance between the two technological ontologies I have described thus far is similar to the distance that separates me from Amerindian knowledge. My access to the ancestral knowledge of the Amazon has almost always been from a distance, an exercise of relational telepathy that invites me to seek through the words of those I read and converse with. Although I traveled to Leticia (the capital of the department of Amazonas in Colombia) for the first time in mid-2022, my relationship with this territory has been mediated to a great extent by my sister Camila Pacheco, anthropologist and designer, my friend Maytik Avirama, human ecologist and sound artist, and my friend Bárbara Santos, an artist whose work is a fundamental part of this and other essays I've written. My relationship with the philosophical thought of the Amazon has been from a distance, mediated by the affection I share with these three women whom I admire so much. The telepathy that interests me does not rely solely on digital devices, but also on conversations, relationships, and experiences that are transmuted to allow us to sense our connection with places we do not directly inhabit. To confront today's ecological crisis, telepathic perspectivism will become increasingly important, as it generates spaces for listening to that which is distant and different to us, which do not prompt us to dominate.

One of the axes of Amazonian ontology is the capacity to inhabit the multiple perspectives that make up the same territory. In this way, perspectivism affirms difference as the central axis of the recognition of the other, extending this human quality to all the actors in a particular territory (Viveiros de Castro, 2014). The relationship between art, technology, and anthropology allows for the possibility of seeing the same territories and bodies from multiple perspectives. In art we can imagine technologies beyond extraction, guided by intuition as we weave bridges between multiple ways of feeling, thinking, and communicating at a distance. In sum, what I am proposing as telepathic perspectivism is very close to artistic thinking and dreaming, allies of the cosmic waters that make up life. It is the set of portals that peer into the abyss between radically different realities. Telepathic perspectivism opens a humid portal to approach technology as a relational and more-than-human performance, which allows for the transformation of energy and matter between territories at a distance. In the middle of the night, I abandon the port's lighthouse in search of even darker waters.

Water and Energy

The day I was born, the planets aligned around water and technology, opening an electrical portal between the salt in my body and the waters that permeate it. According to Hellenistic astrology, my sun is in Cancer, the crab ruled by the moon and the ocean, two material bodies in constant tension. My moon is in Aquarius, the dolphin that inhabits electric air waves to make room for technology and the unconventional. In the space between the sun and the moon, the day and the hour I was born manifest a triangle between my body, water, and electricity. From this cosmic point of view, it is not by chance that almost all of my research in recent years has delved into the relationship between water and technology. All our thoughts and intuitions share the same energy of atoms and celestial bodies. We are vehicles and agents of the deep entanglements between the material and the ineffable, between matter and energy. As I arrive to this new port where I find my astral body, the future of water reveals itself as the future of our bodies, just as the future of our bodies manifests itself in the time of water. To think that technology allows us to detach ourselves from our corporeality is not only a fatal error, but also an outright fallacy.

Through texts, videos, sounds, web projects, and collaborative laboratories, my work since 2016 has immersed me in the poetic and material relationships between water and technology, seeking to weave new narratives to navigate technological waves from an ecological and transhistorical awareness. From the words we use to understand the digital world—such as cloud, torrent, streaming, surfing, sailing—to the undersea cables that enable the global flow of information, the internet bears a close relationship to ocean waters on multiple levels. In addition to being a container for much of the internet's infrastructure, the ocean is also an energetic medium that enables the flow of information on a planetary scale. Due to its high salinity, the sea floor provides a perfect grounding pole for the electrical current that amplifies the signal flowing through undersea cables. Our emails, video calls, images, and sounds are not only contained by ocean water, but depend on it to operate. As Nicole Starosielski comments, "after power crosses the cable, it is routed to an ocean ground bed that grounds intercontinental currents; the ocean completes the cable circuit" (2015, p. 201). The internet's undersea network relies entirely on the electrical conductivity of the aquatic environment, enhanced by the salinity of the seas. As Amerindian cultures have known since time immemorial, salt is a crystal that enables the communication between multiple forms of life.

Figure 13.3. *Still from the video essay The Blue Dot by Juan Pablo Pacheco Bejarano, which speculates on the ecological and epistemological networks woven around the internet. The blue dot represents a real time sign of users' interaction with digital servers, but it also becomes a visual and material metaphor for planet Earth, the world's oceans, and the internet as a humid technology.*

Some recent Microsoft experiments also seek to take advantage of the physical properties of the ocean, designing data centers that use the low temperatures of the seabed to cool the servers. In this way, Silicon Valley companies seek to save on the cost of energy used to lower the high temperatures produced by data processing. As global temperatures continue to rise, I foresee more and more data centers submerged in the ocean depths, further heating the seas in order to continue the expansion of the computer network. An infinite web of catastrophes unleashed by the insistence of technological progress. This utilitarian way of harnessing water as an energy medium, which allows both electrical conductivity and heat dissipation, are part of a long genealogy. The power of water has been the basis of multiple technological systems that have enhanced the colonization of Western modernity over all other cosmotechnics, from the hydraulic mill to hydroelectric dams.

Unlike these extractive systems that reduce water to an inert substance full of potential energy, recognizing the role of water as container and medium allows me to approach telepathy from a relational dimension, bringing to the surface the question of proximity. To humidify telepathy is, then, to see what is in-between rather than ignoring it and trying to eliminate it. Electrical impulses, one of the many forms in which planetary consciousness manifests, cannot be separated from particular territories and their material infrastructures. Submarine cables, data centers, windmills, and hydroelectric plants are bodies linked to the body of the earth and to our own bodies. Downey reminds us that "electromagnetic energy is a river of undulating material. This radiant nature is shared by thoughts, artificial intelligence and video, and explains the very life of the Universe we inhabit" (2013a, p. 265). My proposal is that we immerse ourselves and swim in this consciousness, in its multiple volumes. This way of approaching water allows us to reimagine how we contain it and how it contains us, and thus rescue the possibility of caring for what is distant and different from us.

Deep and expanded reflection with water is essential to reimagine and reshape our relationship with technology. In principle, because water is essential: it is the substance that keeps life in constant movement and transmutation. Water allows us to feel the interconnectedness between every organic and inorganic body as wet matter. From the osmosis that permits the exchange of solvents between cells to the water cycle that connects distant territories, water sets in motion the metabolisms of energetic and material exchange that make up the planetary biosphere. The wet relations of amniotic, salivary, rainy, tropical, and stagnant liquids are part of this telepathic dimension of water, a technology of connection with otherness at a distance. Water—be it liquid, solid or gaseous—is an archive of life, an intercontinental, interplanetary, and intertemporal telepathic communication system. Water is an intelligent network, a constantly transforming source of knowledge, transforming geological formations from the highest mountains to the deepest trenches at the bottom of the ocean.

Water also organizes the world as a voluminous field: up and down, surface and depth, back and forth, right and left, inside and outside. In the ocean, the configuration of surface and depth are in constant flux. The one becomes the other in a continuous intensity of motion. Depth rises to the surface only to descend again; surface submerges and becomes depth. In this sense, water connects us to other geological layers and to the passage of time: a spatial and temporal connector. The wet ontology proposed by Phillip Steinberg and Kimberly Peters rightly invites us to embrace the depth and volume of the ocean as crucial elements to destabilize fixed and solid categories, from the ways in which we are implicated with multiple bodies and systems (Steinberg and Peters, 2015).

More than wet, humid refers to that in-between space between solid and liquid. Although the artist Roy Ascott spoke in the 1990s about moist media to refer to biotechnology, the humidity I speak of is closer to that felt in the tropics; that sticky sensation on the skin when we are in tropical forests that invites us to engage with our surroundings. Tropicalizing technology is, then, recognizing the sticky and humid encounter between bodies at a distance, a sensation that goes beyond the desire to understand. Humid telepathy is that encounter of love that does not apprehend, that appreciates but does not dominate. It is the listening to the sensory noise of our environment. When we allow ourselves to enter into contact with water in a conscious and present way, an inexhaustible source of information is activated.

In most genesis myths, life originates in water. In many Amerindian cultures, and in some of the most recent research from modern science, water is considered capable of retaining this vital memory, like a living archive. The wisdom-keepers of the ancestral cultures of the Amazon follow a rigorous ecological calendar to enter into dialogue with the memory of the rivers.[5] Depending on the time of the year and the day, the river tells different stories about the origin of the world, the state of its headwaters and mouths, and how to organize work for the coming year. The knowledge of each knower is specific to their territorial context, and at the same time it allows them to connect with the spaces that water travels through on a planetary level. Although the experience of a knower in the Amazon is radically different from that of those of us who inhabit the modernized and urban world, the Amerindian cosmovision teaches us that water is not a passive material to be exploited or protected; water is an active agent with which we can commune. From this perspective, mediated by our bodies, we can begin to draw other kinds of maps of the forms of marginality and belonging that emerge from the common waters that we share with the planet and its multiple forms of life (Neimanis, 2017).

The speed at which this wet telepathy operates is different from the speed of digital telepathy. In a way, it is slower and more voluminous, similar to the flow of time to which water currents refer us to. Rivers, streams, and oceans are not homogeneous bodies of water; they are composed of multiple currents of different temperatures and intensities, constantly converging in a nonlinear fashion. In this sense, accessing the living archive of water is closer to an exercise in correspondences, reminiscent of Denise Ferreira da Silva's invitation to investigate and write from the depth, breadth, length, and time of events: "the images of poetic thought are not linear (transparent, abstract, glassy and determined), but fractal (immanent, scalar, abundant and indeterminate), like most of what exists in the world" (2016). Humid telepathy speaks from a nonlinear and fractal perspective, embracing the depth of the ocean and the multivalence of water currents.

Unlike the Platonic idea that reality emanates as a shadow from the sun, I believe that reality emanates from the shadows of the watery depths pulled by the moon's embrace. The humid, pressurized, dark depths of the ocean radiate a wet energy into the world, filling it with water and thus with life. This is the cosmology of the underworld, the place of the high priestess, of the fungi and their infinite cycles. The origin is not a sun; it is a telepathic force from the depths of the sea. This sea is, however, as external as it is internal, similar to Flusser's abyss where the depths of the ocean are indistinguishable from the depths of ourselves (Flusser and Bec, 2012, p. 69). Telepathy is this exercise of knowing the distant in order to know the near; of seeing ourselves reflected in the world as an erotic mirror that is impossible to colonize and exploit. As I leave this astral and energetic port, I abandon my ship and jump into the sea.

Becoming a Dolphin

When I am far from Colombia, I often dream that I become a dolphin that travels through the Atlantic Ocean until I find the mouth of the Magdalena River. Then, at the height of Buenavista, I go up the Río Negro until I find the Tobia River and from there up the San Juan River. There, in San Juan de La Vega where I partially grew up, an always humid and fertile land, lives half of my imagination. My imagination travels at the speed of water currents that, like time, are multiple and voluminous, populated by currents of all temperatures, enveloping each other: up and down, back and forth, inside and outside. Colombia is a colonial fiction that designates an amphibious territory, traversed by thousands of waters that humidify the mountains as they flow down to the sea. The more we inhabit these dreamlike currents, I foresee that we will slowly become dolphins, those mammals that, after tasting the dryness of the land, decided to return to the sea and the river. Abandoning all ports, each one of us will have to find the depth and salinity of our own ocean. Every end is a departure.

In the space between land and water, our transformation will begin with a sensation of vertigo as our vestibular apparatus adapts to a new voluminous coordination, like sailors returning from the sea. The neuroplasticity that Catherine Malabou speaks of in an interview with Hans Ulrich Obrist (2021) will give way to an ever larger cerebellum, like that of dolphins, to allow us to find our balance in deep water. Then, our eyes will begin to distort our perception of space, like rippling landscapes reflected in water surfaces. We will see our distorted reflection, widening the portal to our depths. When we complete our metamorphosis, we will perceive with greater amplitude the frequencies that escape our sensory field. The water will be our new screen, a border and a portal at the same time. Becoming a dolphin will be the way in which water's life archive will pass through our bodies and become consciousness.

Inspired by my friend and artist Mariana Murcia, in the last few months I have bathed in icy waters in The Netherlands and Spain. After some breathing exercises that warm up the core of the body, entering icy waters is a discharge of memories. The muscles tense and the atomic bonds of the skin are activated, as if an electric wave were running through the membrane that separates us from what is external and distant. The first time I bathed in the Maas River at the end of November of 2022, I was invaded by the memory of the electricity that gave rise to life in the sea, molecules excited by deep temperature changes. From her deep relationship with the ancestral knowledge of the Amazon, Bárbara Santos reflects that "Western culture, by delegating memory to books, computers, photographs, etc., does not know how to activate memory through the body" (Santos, 2019, p. 82). Through eating, singing and dancing, many Indigenous knowers activate the electric memories of deep times contained in the body. In a parallel world, when I swim in the icy waters of the Maas and of the Navalmedio dam in Spain I activate a memory of the origin of life. Technology is, fundamentally, this electric sea that flows through the membranes of our bodies.

Similar to the membranes of undersea cables, water is also a fundamental communication technology for living organisms. Through osmosis, cells in our body tissues exchange salts and molecules across their semipermeable membranes. These membranes enabled the emergence of complex life in the ocean millions of years ago, driving the intertwined dependence between information and media. One unicellular bacterium engulfs the other and thus creates complex life. Multicellular life is the result of cannibalism between membranes, insisting on surviving inside each other in the vast Precambrian sea (Margulis, 1997). Membranes are an artificial intelligence that creates the very notion of an inside and an outside, spaces where both difference and similarity emerge. In an interview with Bárbara Santos, anthropologist Stephen Hugh Jones suggests that the body is a structure of tubes: "we are all tubes and to this extent our body is a tube containing other tubes in which there are openings: I speak through my mouth, I reproduce in the womb" (quoted in Santos, 2019, p. 82). Between tubes, cables, and membranes, the body of the internet and our bodies are reflections of the same territory contained by planetary waters.

So what rituals allow us to cross the material and semiotic membranes between ourselves, our environments, and other species? At what point does the subject begin and the environment end? In discussing the relationship between telepathy, architecture, and video, Juan Downey reminds us that the screen is also that pulsating refuge of exchange, a membrane between inner and outer myths (Downey, 2013a, p. 264). The screen and the surface of water, intermediate membranes, allow for constant mediation between the near and the distant. In this sense, wet telepathy reminds us that to know anything is to know oneself (Agamben, 1995). Beyond grasping and understanding, humid telepathy invites us to immerse ourselves in the medium, and to love without the desire to decipher the other. The humid is human, and it is humus; the earth from which the possibility of growing and caring arises. In this fertile space of interpretation, emerges the possibility of relating beyond extraction.

Every time I use a hydrophone to listen to the dolphins' home, I enter into a multitemporal connection with water currents. The hydrophone transduces the vibrations within the water into electrical impulses, making it possible to listen and record beneath its surface. This transduction through media is fundamental to telepathy, which proposes a dialogue between spaces and agents in a game of semipermeable messages (Helmreich, 2007). When I stop and listen to the clamor of the water flowing down from the *páramo*,[6] I begin to feel like an agglomeration of spaces submerged in time; a continuous flow; a sea of events passing at different speeds. In the midst of the electric static, I begin to perceive the agency of other lifeworlds and to tune in to the multiple forces that shape our world and that do not depend on human will. The future of water will be to break down the walls that contain it, to overflow the spouts of cities and hydroelectric dams, and to humidify the wetlands withered by reason.

Notes

[1] Thanks to Bruno Alves de Almeida for highlighting this perspectival phenomenon during our joint research on the seabed during the PACT Zollverein residency in Germany, 2021.

[2] To learn more about the link between colonial routes and submarine cable routes, you can consult my research on Atlantis-2, the first submarine internet cable between Europe and South America (Pacheco Bejarano, 2021).

---

[3] In another essay I have discussed this issue more extensively, by proposing a difference between extractive telepathy and regenerative telepathy (Pacheco Bejarano, 2020).

[4] The coup d'état to the democratically elected government of Salvador Allende marked a critical moment in Latin American geopolitics, as it was supported by the CIA in order to install the military man Augusto Pinochet in power and thus ensure the creation of a neoliberal experiment. According to many analysts, Pinochet's regime inaugurated the neoliberal policies that were imposed ironclad during the 1980s by the US and British governments globally and stifled the incursion of cybernetics into the socialist model applied by Allende (Medina, 2011).

[5] This reflection arises from a conversation with Bárbara Santos several years ago, where she shared with me an interview with her friend Fabio Valencia, a Makuna knower and leader in the Pirá Paraná River area in the Colombian Amazon.

[6] Páramos are highland moors found in the Andes, tropical and cold ecosystems where plants have specialized to capture water from the air's moisture.

References

Agamben, G. (1995). *Idea of prose*. Albany, NY: State University of New York Press.

Downey, J. (2013a). Architecture, video, telepathy. In: González, J., Ramírez Moyao, A. (Eds.), *Juan Downey. Una Utopía de La Comunicación/ A Communications Utopia*, pp. 260-267. Mexico D.F.: Fundación Olga y Rufino Tamayo.

Downey, J. (2013b). Descriptive Accounts of Trans Americas Video, 1973-1975. In: González, J. Ramírez Moyao, A. (Eds.), *Juan Downey. Una Utopía de La Comunicación/A Communications Utopia*, pp. 270-94. Mexico D.F. : Fundación Olga y Rufino Tamayo.

Downey, J. (2013c). Technology and beyond. In: González, J., Ramírez Moyao, A. (Eds.), *Juan Downey. Una Utopía de La Comunicación/ A Communications Utopia*, pp. 252-255. Mexico D.F.: Fundación Olga y Rufino Tamayo.

Ferreira da Silva, D. (2016). Fractal Thinking. In: Baierl L., Bu Shea S., Elifritz P. (Eds.), *About, Search, Archive*, (2), Available at: https://accessions.org/article2/fractal-thinking/

Flusser, V. (2014). *Gestures*. Minneapolis: University of Minnesota Press.

Flusser, V., Bec, L. (2012). *Vampyroteuthis Infernalis*. Minneapolis: University of Minnesota Press.

González, J. (2013). A communication utopia by Juan Downey. In: González, J., Ramírez Moyao, A. (Eds.), *Juan Downey. Una Utopía de La Comunicación/ A Communications Utopia*, pp. 10–81. Mexico D.F. : Fundación Olga y Rufino Tamayo.

Guagnini, N. (2008). Feedback in the Amazon. *October Magazine*, 125, pp. 91-116.

Helmreich, S. (2007). An Anthropologist Underwater: Immersive Soundscapes, Submarine Cyborgs, and Transductive Ethnography. *American Ethnologist*, 34(4), pp. 621-41.

Hui, Y. (2021). *Art and Cosmotechnics*. Minneapolis: University of Minnesota Press.

Malabou, C., Obrist, H. U. (2021). Plasticity, Intelligence and Mind. In: Vickers, B., Allado-McDowell, K (Eds.), *Atlas of Anomalous AI*. London: Ignota Books.

Margulis, L. (1997). From Kefir to Death. In: Margulis, L., Sagan, D. (Eds.), *Slanted Truths: Essays on Gaia, Symbiosis and Evolution*, New York: Springer-Verlag.

Medina, E. (2011). *Cybernetic Revolutionaries: Technology and Politics in Allende's Chile*. Cambridge MA: MIT Press.

Mignolo, W.D. (2011). *The Darker Side of Western Modernity: Global Futures, Decolonial Options*. Durham NC: Duke University Press.

Neimanis, A. (2017). *Bodies of Water: Posthuman Feminist Phenomenology*. London: Bloomsbury Academic.

Pacheco Bejarano, J.P. (2021). Ruins across the Atlantic: speculations on the colonial and mythological genealogies of the internet's submarine infrastructure, *Proceedings of Politics of the Machines - Rogue Research 2021*, ScienceOpen, pp. 138-144. Available at: https://doi.org/10.14236/ewic/POM2021.18

Pacheco Bejarano, J.P. (2020). Telepathy without the internet. *Dispatches: Journal of Visual Culture*, pp.1-10. https://www.academia.edu/44381152/Telepathy_Without_the_Internet

Parikka, J. (2015). *A Geology of Media*. Minneapolis: University of Minnesota Press.

Santos, B. (2019). *Curación como Tecnología: Basado en Entrevistas a Sabedores de la Amazonía*. [Healing as Technology: Based on Interviews with Knowers from the Amazon]. Bogotá: Idartes.

Star, S.L., Bowker, G.C. (2006). How to Infrastructure. In: Lievrouw, L.A, Livingstone, S. (Eds.), *Handbook of New Media: Social Shaping and Social Consequences of ICTs*, pp. 230-245. London: Sage Publications.

Steinberg, P., Peters, K. (2015). Wet Ontologies, Fluid Spaces: Giving Depth to Volume through Oceanic Thinking. *Environment and Planning D: Society and Space*, 33 (2), pp.247-64. Available at: https://doi.org/10.1068/d14148p.

Starosielski, N. (2015). *The Undersea Network*. Durham NC: Duke University Press.

Viveiros de Castro, E. (2014). *Cannibal Metaphysics: For a Post-Structural Anthropology*. Minneapolis: Univocal.

Zuboff, S. (2019). *The Age of Surveillance Capitalism: The Fight for a Human Future at the New Frontier of Power*. London: Profile Books.

www.ingramcontent.com/pod-product-compliance
Lightning Source LLC
Chambersburg PA
CBHW060247230326
41458CB00094B/1475

* 9 7 8 1 8 0 4 4 1 0 6 8 4 *